Reliability Design of Mechanical Systems

Seongwoo Woo

Reliability Design of Mechanical Systems

A Guide for Mechanical and Civil Engineers

 Springer

Seongwoo Woo
Reliability Association of Korea
Seoul
Korea (Republic of)

ISBN 978-3-319-50828-3   ISBN 978-3-319-50829-0 (eBook)
DOI 10.1007/978-3-319-50829-0

Library of Congress Control Number: 2016960023

© Springer International Publishing AG 2017

This work is subject to copyright. All rights are reserved by the Publisher, whether the whole or part of the material is concerned, specifically the rights of translation, reprinting, reuse of illustrations, recitation, broadcasting, reproduction on microfilms or in any other physical way, and transmission or information storage and retrieval, electronic adaptation, computer software, or by similar or dissimilar methodology now known or hereafter developed.

The use of general descriptive names, registered names, trademarks, service marks, etc. in this publication does not imply, even in the absence of a specific statement, that such names are exempt from the relevant protective laws and regulations and therefore free for general use.

The publisher, the authors and the editors are safe to assume that the advice and information in this book are believed to be true and accurate at the date of publication. Neither the publisher nor the authors or the editors give a warranty, express or implied, with respect to the material contained herein or for any errors or omissions that may have been made.

Printed on acid-free paper

This Springer imprint is published by Springer Nature
The registered company is Springer International Publishing AG
The registered company address is: Gewerbestrasse 11, 6330 Cham, Switzerland

Preface

In the beginning of the twentieth century, new sophisticated mechanical systems such as bridges, rockets, automobiles, airplanes, and space shuttles were designed and built for people to live comfortable lives through the engineering design processes. Typical design process can be broadly summarized as (1) define the problems, (2) develop the product–prototype, design and testing, (3) production. Due to the frequent occurrence of disasters for new products, product reliability has become one of increasingly important factors (to consider) because of cost, competition, public demand, and adaptation of new technology. The most effective way to protect the reliability disaster is to develop the reliability-embedded design process including its methodology in parallel with the established design process.

As products with multiple modules require higher performance and material cost reduction, the reliability design of product has become more complex and increases the risk of product failure. The studies of reliability engineering have been deepened to prevent the reliability disasters of the past century. Even though there are a large number of concepts, theory, and texts on reliability, an up-to-date book for emphasizing the new methodology of reliability design is still required to prevent the reliability disasters of the mechanical/civil system.

From the standpoint of economics, company will decrease the operation profit for a failure in its expected product lifetime because of Product Liability Law in the global market. All products from tires to electric components are fabricated from the structure (or materials) that will tend to degrade or break down abruptly by random loads. The mechanical system can eventually fracture due to fatigue which can result from cyclical stresses (or loads). When products are subjected to random loads, they start the void in material (or design defects), propagate, and rupture it. If failure for a new product happens, the product may no longer meet the established specifications for proper product functionality. To avoid product failure in lifetime, product should be designed to robustly withstand a variety of loads.

The main objectives of writing this book are focused on explaining the development necessity of the reliability-embedded design process and its methodology.

As reliability methodology, we will suggest the new parametric accelerated life testing (ALT) that meets those market requirements—higher performance, material cost reduction, and higher reliability in field. The reliability-embedded design process consists of parametric ALT plan, failure mechanism and design, acceleration factor, sample size equation, and the parametric ALT. It produces the reliability quantitative test specifications (RQ) in accordance with the reliability target. A parametric ALT method therefore will assess the reliability of product subjected to repetitive stresses.

Based on the market data, parametric ALT plan will set up the reliability target of product and its modules. Mechanical system in field subjected to loads arise how to design product for the failure mechanisms—fatigue and fracture. The accumulated damage in system like palmer miner rule can be represented at the time-to-failure model. The acceleration factor with a new effort concept (or loads) was derived from a generalized life-stress failure model. So the new sample size equation with the acceleration factor enabled the parametric ALT to quickly evaluate the expected lifetime of product. This parametric ALT should help an engineer to uncover the missing design parameters affecting reliability during the design process of new product.

Consequently, if applied in the established design process, new parametric ALT helps companies to improve new product reliability and avoid the recalls of product failures in field. As the improper design parameters in the design phase are identified by this reliability design method, the product will improve the reliability that will be measured by the increase in lifetime, L_B, and the reduction in failure rate, λ. Product will meet the reliability target in industry. This book will help to prevent the reliability disaster through the parametric ALT. We also provide a lot of parametric ALT examples that are effective to be understood in the mechanical/civil field.

This book is composed of nine chapters. Chapter 1 presents the present aspect and need of reliability engineering in the advance of modern technology. Chapter 2 reviews the historical reliability disasters and their root cause within the past century. It will explain the significance of reliability assessment, and its methodology need to prevent reliability disasters in the design process. Chapter 3 will explain the most important fundamental definitions of statistics and probability theory, the mathematical essentials of reliability engineering, and the most significant aspects of reliability engineering developed within the past century. It will help one to understand the basic concepts of reliability methodologies that will be discussed in Chap. 8. Chapter 4 through Chap. 6 present load analysis, stress concept, and a brief overview of the typical reliability failure mechanism of product—fatigues and fractures. Chapter 7 will present the fundamental concepts of the parametric ALT in product that will be the core of this book. Chapter 8 will also present case studies that are useful in a variety of engineering areas. Chapter 9 will cover the future aspects of parametric ALT in mechanical product that will be developed as system engineering.

This book is intended to introduce the prerequisite concepts of the parametric ALT for senior level undergraduate and graduate students, professional engineers, college and university level lecturers, researchers, and design managers of the engineering system. We hope this noble methodology explained in this book will help to prevent the reliability disasters of new product in field. The authors would also like to thank Springer for the publishing of this work, especially Mayra Castro, Springer DE. With their help, this book has been published.

Seoul, Korea (Republic of)   Seongwoo Woo

Contents

1 Introduction to Reliability Design of Mechanical/Civil System 1
   1.1 Introduction 1

2 Reliability Disasters and Its Assessment Significance 7
   2.1 Introduction 7
   2.2 Reliability Disasters 10
      2.2.1 Versailles Rail Accident in 1842.................... 12
      2.2.2 Tacoma Narrows Bridge in 1940 13
      2.2.3 De Havilland DH 106 Comet in 1953................. 14
      2.2.4 G Company and M Company Rotary Compressor Recall in 1981... 15
      2.2.5 Firestone and Ford Tire in 2000 17
      2.2.6 Toshiba Satellite Notebook and Battery Overheating Problem in 2007 18
      2.2.7 Toyota Motor Recalls in 2009 19
   2.3 Development of Reliability Methodologies in History 20
      2.3.1 In the Early of 20s Century—Starting Reliability Studies................................ 20
      2.3.2 In the World War II—New Electronics Failure in Military....................................... 24
      2.3.3 In the End of World War II and 1950s—Starting the Reliability Engineering 26
      2.3.4 In the 1960s and Present: Mature of Reliability Methodology—Physics of Failure (PoF) 30
   References... 34

3 Modern Definitions in Reliability Engineering 35
   3.1 Introduction 35
      3.1.1 Bathtub Curve................................... 36
   3.2 Fundamentals in Probability Theory........................ 37
      3.2.1 Probability...................................... 38
      3.2.2 Probability Distributions 40

	3.3	Reliability Lifetime Metrics	44
		3.3.1 Mean Time to Failure (MTTF)	44
		3.3.2 Mean Time Between Failure (MTBF)	45
		3.3.3 Mean Time to Repair (MTTR)	46
		3.3.4 BX% Life	46
		3.3.5 The Inadequacy of the MTTF (or MTBF) and the Alternative Metric BX Life	47
	3.4	Statistical Distributions	49
		3.4.1 Poisson Distributions	49
		3.4.2 Exponential Distributions	51
	3.5	Weibull Distributions and Its Applications	52
		3.5.1 Introduction	52
		3.5.2 Shape Parameters β	54
		3.5.3 Confidence Interval	54
		3.5.4 A Plotting Method on Weibull Probability Paper	55
		3.5.5 Probability Plotting for the Weibull Distribution	56
	Reference		59
4	**Failure Mechanics, Design, and Reliability Testing**		**61**
	4.1	Introduction	61
	4.2	Failure Mechanics and Designs	63
		4.2.1 Product Design—Intended Functions	64
		4.2.2 Specified Design Lifetime	66
		4.2.3 Dimensional Differences Between Quality Defects and Failures	67
		4.2.4 Classification of Failures	68
	4.3	Failure Mode and Effect Analysis (FMEA)	70
		4.3.1 Introduction	70
		4.3.2 Types of FMEA	72
		4.3.3 System-Level FMEA	72
		4.3.4 Design-Level FMEA	73
		4.3.5 Process-Level FMEA	73
		4.3.6 Steps for Performing FMEA	74
	4.4	Fault Tree Analysis (FTA)	79
		4.4.1 Concept of FTA	79
		4.4.2 Reliability Evaluation of Standard Configuration	83
	4.5	Robust Design (or Taguchi Methods)	85
		4.5.1 A Specific Loss Function	86
		4.5.2 Robust Design Process	89
		4.5.3 Parameter (Measure) Design	90
		4.5.4 Tolerance Design	90
		4.5.5 A Parameter Diagram (P-Diagram)	91
		4.5.6 Taguchi's Design of Experiment (DOE)	91
		4.5.7 Inefficiencies of Taguchi's Designs	93

	4.6	Reliability Testing	94
		4.6.1 Introduction	94
		4.6.2 Maximum Likelihood Estimation	95
		4.6.3 Time-to-Failure Models	97
		4.6.4 Reliability Testing	100
5	**Load Analysis**		**107**
	5.1	Introduction	107
	5.2	Modeling of Mechanical System	108
		5.2.1 Introduction	108
		5.2.2 D'Alembert's Modeling for Automobile	109
	5.3	Bond Graph Modeling	112
		5.3.1 Introduction	112
		5.3.2 Basic Elements, Energy Relations, and Causality of Bond Graph	113
		5.3.3 Case Study: Hydrostatic Transmission (HST) in Seaborne Winch	118
		5.3.4 Case Study: Failure Analysis and Redesign of a Helix Upper Dispenser	124
	5.4	Load Spectrum and Rain-Flow Counting	127
		5.4.1 Introduction	127
		5.4.2 Rain-Flow Counting	129
		5.4.3 Goodman Relation	131
		5.4.4 Palmgren-Miner's Law for Cumulative Damage	132
	References		137
6	**Mechanical System Failures**		**139**
	6.1	Introduction	139
	6.2	Mechanism of Slip	142
	6.3	Facture Failure	144
	6.4	Fatigue Failure	146
		6.4.1 Introduction	146
		6.4.2 Type of Fatigue Loading	147
		6.4.3 Stress Concentration at Crack Tip	150
		6.4.4 Crack Propagation and Fracture Toughness	152
		6.4.5 Crack Growth Rates	153
		6.4.6 Ductile–Brittle Transition Temperature (DBTT)	155
		6.4.7 Fatigue Analysis	157
	6.5	Stress–Strength Analysis	159
	6.6	Failure Analysis	160
		6.6.1 Introduction	160
		6.6.2 Procedure of Failure Analysis	162
		6.6.3 Case Study: PAS (Photo Angle Sensor) in Automobile	164
		6.6.4 Fracture Faces of Product Subjected to a Variety of Loads in Fields	167
	References		169

7 Parametric Accelerated Life Testing in Mechanical/Civil System ... 171

- 7.1 Introduction 171
- 7.2 Reliability Design in Mechanical System 172
- 7.3 Reliability Block Diagram and Its Connection in Product 175
- 7.4 Reliability Allocation of Product 176
  - 7.4.1 Introduction 176
  - 7.4.2 Reliability Allocation of the Product 177
  - 7.4.3 Product Breakdown 178
- 7.5 Failure Mechanics, Design, and Reliability Testing 184
- 7.6 Parametric Accelerated Life Testing 187
  - 7.6.1 Acceleration Factor (AF) 188
  - 7.6.2 Derivation of General Sample Size Equation 193
  - 7.6.3 Derivation of Approximate Sample Size Equation 196
- 7.7 The Reliability Design of Mechanical System and Its Verification 198
  - 7.7.1 Introduction 198
  - 7.7.2 Reliability Quantitative (RQ) Specifications 200
  - 7.7.3 Conceptual Framework of Specifications for Quality Assurance 204
- 7.8 Testing Equipment for Quality and Reliability 206
  - 7.8.1 Introduction 206
  - 7.8.2 Procedure of Testing Equipment Development (Example: Solenoid Valve Tester) 209
- References 218

8 Parametric ALT and Its Case Studies 221

- 8.1 Failure Analysis and Redesign of Ice Maker 221
- 8.2 Residential Sized Refrigerators During Transportation 229
- 8.3 Water Dispenser Lever in a Refrigerator 233
- 8.4 Refrigerator Compressor Subjected to Repetitive Loads 242
- 8.5 Hinge Kit System (HKS) in a Kimchi Refrigerator 253
- 8.6 Refrigerator Drawer System 263
- 8.7 Compressor Suction Reed Valve 268
- 8.8 Failure Analysis and Redesign of the Evaporator Tubing 279
- 8.9 Compressor with Redesigned Rotor and Stator 288
- 8.10 French Refrigerator Drawer System 296

9 Parametric ALT: A Powerful Tool for Future Engineering Development 307

- Reference 310

Chapter 1
Introduction to Reliability Design of Mechanical/Civil System

Abstract From the standpoint of system engineering, this chapter will be discussed with the necessities of new reliability-embedded design process to catch up with incredible technology innovations for new product. The modern products should survive to compete other products in global. They are often required to have higher performance and reliability for the necessary intended functions, though the product cost and developing time has to reduce it gradually. Because the new product hardly satisfies the requirements within the limited development time, there is the presence of risks on reliability disasters at all times. New assessment methodology of reliability in the reliability-embedded design process is required to protect the massive recall in lifetime. It will therefore be possible to discover the design defects in the design phase through new reliability methodology that could enable product to satisfy the reliability target of product.

Keywords Reliability-embedded design process · Reliability testing methodology · Design requirements · Performance · Cost-down

1.1 Introduction

As the frequent recalls for the new product occur globally, the term of product reliability seems to friendly be used to everyday life. The product quality and its reliability seem to become important requisites to ensure the continued success in the current global competitive marketplace. If the customer does not satisfy the product quality, it will be expelled in global market. Thus, it is important for the product design team to understand customer expectations or voices. The product reliability is to create a product that can properly work the required intended functions under all environmental conditions in operation lifetime of product. To achieve the reliability requirements in field, numerous concepts—bathtub, MTBF, and failure rate have been established in the last century. They also require the fundamental knowledge of the probability and statistics. As they are put to use, reliability could quantitatively describe the failure data from the marketplace.

As seen in Fig. 1.1, the product development in the field of the mechanical/civil engineering system is continually confronting to be satisfied with the end user's requirements—high control performance, high response, energy efficiency, low noise, high reliability, long life time, the latest hardware design, contamination resistance, low price, compact, and highly portable weight, and precision control for wide frequency range. To survive the competitive global environments, the company should manufacture the high-performance products that meet the customer expectations or their specifications.

Engineers, however, wonders if product development satisfies the requirement of these attributes in reality. Ironically, to get those attributes in the product design such as automotive and cell phones, the product development times are continually decreasing. On the other hands, product reliability in marketplace is highly required due to the recall costs. Thus, new product is hard to match the market requirements of product—cost reduction, the shortening developing time, higher performance and reliability. From the standpoint of system engineering, companies are asked to establish the design process of satisfying the product requirements in the short development time. For example, while product development time—automobile continued to shorten from 65 months to 24 months, reliability required increases from 0.9 to 0.99. These declines mean that companies have reliability methodology tools closely tied to the development process itself (Fig. 1.2).

As market is requiring, a myriad of technology innovations are constantly emerging and disappearing. People owing to a state-of-the-art technical renovations broaden their lives and widen their boundaries. On the other hand, they also experience frequent malfunctions as new product has been released in the marketplace. They ask to replace the problematic product with new one. To satisfy the end user's requirements, most global companies have to be established in the product developing process that can find out the problematic design. As a

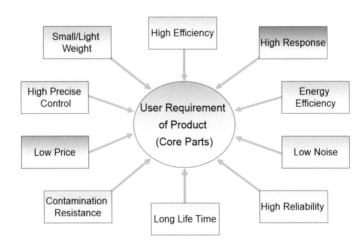

Fig. 1.1 Customer requirements of product (or core parts)

1.1 Introduction

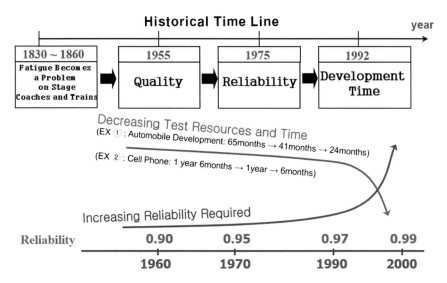

Fig. 1.2 Historical time line for product quality

matter of fact, they have traditional methodologies to achieve high product quality as qualitative-FMEA (Failure Mode Effective Analysis) and FTA (Fault Tree Analysis). But there is no quantitative methodology-reliability testing. And competitive company in high technology industries only can prosper in markets whose customers satisfy extreme needs, such as safety-critical mechanisms (aircraft) or high technology military armaments.

The established developing process of product in company can be largely classified as Research & Development (R&D) and Quality Assurance (QA). R&D is a core part of the modern company because major design decisions in firms are made based on its technical level. As companies define the design requirements from customer needs or past experiences, they start to develop new product that satisfies those specifications. R&D activities also are conducted by departments with high specialized person in technique. They design architectural structures, proper materials, and robust systems while considering the limitations-practicality, regulation, safety, and cost. A professional engineer can apply the scientific methods to solve out engineering problems by FEA (Finite Element Analysis). They also use the advanced manufacturing processes, expensive safety certifications, specialized embedded software, computer-aided design software, electronic designs, and mechanical subsystems. The design process of product embedded in reliability concept can be briefly defined as qualitative design process and quantitative design process. It will briefly flow down the product planning, concept design, basic design, prototype, detail design, and production (See Fig. 1.3).

For a detailed design of product, Quality Assurance (QA) will determine if the product is satisfactory to each company specifications. In other words the quality of product may be explained as the product specifications that are summarized as

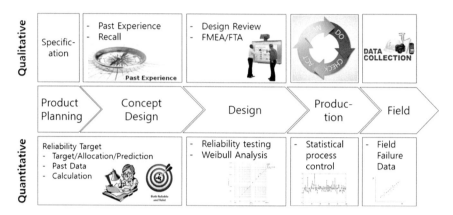

Fig. 1.3 New developing process of product embedded in reliability concept

perception of the degree or the end-user's expectations. Quality verifications in these forms was initially established by National Aeronautics and Space Administration (NASA), the military and nuclear industries in the 1960s. The specification-oriented development process was designed to develop better products that have no design modifications or technical innovations at that time. And it was focused on manufacturing, testing, and quality control, rather than design. At that time the typical design tools—design review, FMEA and FTA are to qualitatively accomplish the specifications of product quality but from a standpoint of quantitative quality, there is no design process and reliability methodology to achieve the reliability targets.

Because the traditional methodology in the design process cannot find the chronic problems for design issues of new technologies, products always have inherent design problems in marketplace that might unceasingly give rise to massive recalls. For instance, the Boeing 787 Dreamliner experienced some problems due to new design elements—the fuselage of carbon-fiber-reinforced plastic (CFRP) and the electrical system incorporating lithium-ion batteries, which ultimately resulted in grounding.

From a standpoint of reliability engineering, why do the historical reliability disasters such as the explosion of challenger happen continually? They might come from the faulty components that have the missing design parameters not found in the design process. The suspected components mounted in product determine the lifetime of product when they are subjected to the wearout stress or overstress under the end user operating or environmental conditions. So to find out the problematic parts mounted in product, new reliability methodology is continually required. New reliability-embedded developing process with new reliability methodology should be suggested in (1) product reliability target/allocation/prediction, (2) reliability testing and Weibull analysis, (3) finding the design problem of the suspected parts, and (4) the analysis of the field failure data.

1.1 Introduction

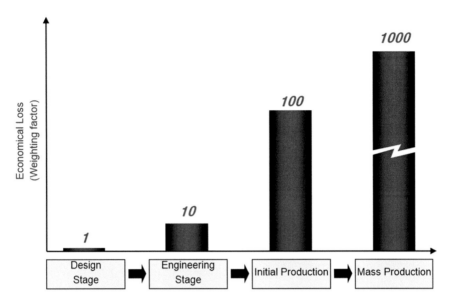

Fig. 1.4 Relationship between failure costs and product life cycles

In the relationship between failure costs and product life cycles, we know that the earlier reliability in the design process is applied, the greater the profit is obtained. Total cost of a product is determined by its design and its value is approximately 70%. For example, if cost $1 is required to rectify a design defect in the prior design stage, the cost would increase to $10 after final engineering stage, $100 at the initial production, and $1000 at the mass product stage (Fig. 1.4).

Before releasing the product to the marketplace, the key factors of quality activities are to discover the missing design parameters that have not been found in the design process. As a result, new quantitative reliability methodologies in the reliability-embedded developing process should search out the problematic components and prevent reliability disasters in the design phase that traditional R&D developing process cannot solve. As Failure Analysis (FA), they should be closely looked in the product design phase that failures in field may happen. In the reliability-embedded process engineer generates reliability quantitative specifications that fit to a newly developed product and increase the lifetime of product by correcting the missing design parameters.

As reliability testing, fatigue failure due to design defects were traditionally assessed from fatigue testing. Fatigue testing is critical, but it has many limitations—(1) requires many physical prototypes, (2) difficult to achieve realistic tests, (3) slow and expensive difficult to conduct early in the design process, (4) requires many tests and statistical interpretation. However, the product still has the inherent design errors because they reveal to use in field soon or late.

Consequently, companies are required to develop new reliability methodology that make up for the weak points and find them in the design process. As reliability quantitative test specifications (RQ), new reliability methodology that will be discussed in Chap. 7 is mainly focused on the fundamental concepts of reliability and parametric ALT. First, after reliability disasters are reviewed, reliability assessment tools developed in history will explain its strengths and weaknesses. We will look over the concepts of failure mechanism, design and reliability testing in the mechanical/civil systems. The parametric ALT in Chap. 7 is a core part of the reliability-embedded product developing process. It consists of parametric ALT plan, failure mechanism and design, acceleration factor, and sample size equation. As a quantitative method, the parameter ALT will be helpful to set up the reliability target and establish the verification specifications over the full range of functions fitted to each product. A variety of parametric ALT case studies in Chap. 8 will also be suggested to clearly understand the methodology of the parametric ALTs.

Chapter 2
Reliability Disasters and Its Assessment Significance

Abstract This chapter will review the historical reliability disasters including natural hazard and the methodology of its reliability assessment that were developed in the last century. Most of reliability disasters come from the complexity of product intended functions due to the customer requirements and its inheritance design defects as new technology introduces. As countermeasures against reliability disasters, methodology of reliability assessment like bathtub curve and Weibull analysis has been developed in the previous century. For instance, the frequent derail accidents of railroad in the early nineteen century started the research for its root cause and made the S–N Curve. The chronic failed vacuum tube in the WWII created the bathtub curve. The FMEA, FTA and Weibull analysis for reliability testing today have been widely used in company as NASA developed for the space shuttle program in the mid-sixties. Now Physics of Failure (POF) become more important tools to analyze the failure mechanics since the introduction of Integrated Circuit (IC), transistor radio and TV in the late 1960s. However, in the field of mechanical/civil system, representative POFs were still fracture and fatigue but not introduced to find the problematic parts by reliability testing method.

Keywords Reliability disasters · Natural hazard · Bathtub curve · Physics of Failure (POF)

2.1 Introduction

A disaster—oil spill, nuclear plant accident and the others is a deep-felt functional failure of the product accompanying catastrophic human, economic, or environmental impacts, which has no predicting ability of the community or society to manage its own resources. Thus, people often have been learning the lessons and setting up countermeasures because they can be prevented if its root causes were known previously. For instance, the RMS Titanic in 1912 had approximately 2200 people on board. It took two hours and forty minutes to sink and drowned to deaths of more than 1500 people. When the crews sighted the iceberg, Titanic was unable to quickly turn and collided the floating ice in right side. There was no plan for rescue, though the ship was sinking fast.

The sinking of the titanic was caused primarily by the brittleness of the steel used to construct the hull of the ship. In the icy water of the Atlantic even a small impact between iceburger and ship could have caused a large amount of damage. The impact of an iceberg on the ship's hull resulted in brittle fracture of the bolts that were holding the steel plates together (Fig. 2.1).

After that, every ship has to have an evacuation plan in danger. When disasters have been studied for more than 40 years, disasters in history might have been seen as the result of inappropriately risk management or mutual combination of hazards and vulnerability. Since most of disasters result from human-made carelessness or proper management measures can partially prevent it from developing into a disaster.

In addition, a natural hazard will briefly be explained before going into the reliability disasters. Developing countries-the Philippines, Nepal are suffering from natural hazards that are caused into more than 95% of all deaths, and are 20 times greater than that of industrialized countries. They are all natural hazards that kill thousands of people and destroy billions of dollars of habitat and property each year. Because there is no single root cause from natural hazards, they are more common in developing countries because those countries have no emergency systems. Typical examples are flood, transport accidents, nuclear explosions/radiation, and an earthquake that causes a tsunami, resulting in coastal flooding.

Fig. 2.1 Sinking picture of the RMS Titanic Illustration for "Die Gartenlaube" magazine by Willy Stöwer, 1912

2.1 Introduction

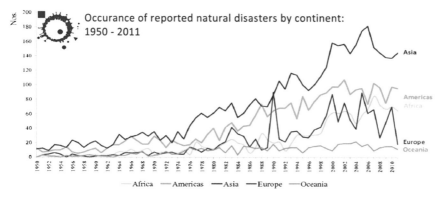

Fig. 2.2 Occurrence of reported natural disasters by continent (1950–2011). *Sources* Reported natural disasters 1950–2011 from CRED

Recorded in magnitude 9.2, the 1964 Alaskan earthquake occurred on March 27 and resulted in 139 deaths. Anchorage experienced great destruction or damage to many houses, buildings, and infrastructures like roads, particularly in the several landslide zones along Knik Arm.

Recently the population growth in the world and its environmental effects has increased the severity of natural hazards due to the global warming, depletion of the ozone layer, and the Central Pacific El Nino phenomenon. The well-known several reasons—the tropical climate and artificial land forms, coupled with the deforestation of the Amazon, unplanned growth proliferation, non-engineered barbaric constructions—in the worldwide make the natural hazard areas more vulnerable (Fig. 2.2).

Typical countermeasures against natural hazards can be classified into (1) research into the scientific aspects of disaster prevention, (2) the strengthening of the disaster prevention system, its facilities and equipment, and other preventive measures, (3) construction projects-dam and meteorological[weather] observations-designed to enhance the country's ability to defend against disasters, (4) and emergency measures and recovery operations. Developing countries suffer chronically from natural hazards, though several preventions for natural hazard.

After the Kobe earthquake in 1981 claimed some 5100 lives, Japan updated its building guidelines, added fresh fuel to another round of research on earthquake safety and disaster management. In 2000, the country's building codes with specific requirements and mandatory checks were revised. From 1979 to 2009, Shizuoka prefecture poured more than $4 billion into improving the safety of hospitals, schools, and social welfare facilities. Though Japanese cities often shake, they rarely topple. Because Japan is located in the Pacific rim, one of the Earth's most violent earthquake and volcano zones, they are still vulnerable.

2.2 Reliability Disasters

Reliability disasters are the consequence of technological risks due to product failures or human-induced damages. Typical examples include transport accidents, industrial accidents, oil spills, and nuclear explosions/radiation. Deliberate terrorism, like the September 11 attacks, may also be put in this category. For example, when the basic cause of reliability disasters in modern product are considered, there might have been product complexity as demanded by customers. Today a typical Boeing 747 jumbo jet airplane is made of approximately 4.5 million parts including fasteners, multiple modules, and subsystems. An automobile is made of more than 25,000 parts, multiple modules, and subsystems. In 1935 a farm tractor was made of 1200 critical parts, and in 1990 the number increased to around 2900. Even for relatively simpler products such as bike, there has been a significant increase in complexity with respect to parts. Consequently, the product design such as automobile is becoming to require these parts to withstand the environmental and user loading conditions (see Fig. 2.3).

Together with product complexity, there are possibilities for the inherent design problems of the parts. A study performed by the U.S. Navy concerning parts failure causes attributed 43% of the failures to design, 30% to operation and maintenance, 20% to manufacturing, and 7% to miscellaneous factors [1]. While the design cost occupies only 5%, the cost influence holds 70%. Thus, we know that quality cost could be saved if the design factors of faulty parts are known before mass production (Fig. 2.4).

Thus we will have suggested typical examples—space shuttle challenger, Chernobyl nuclear reactor, Point Pleasant Bridge, and the others. They also might have been prevented if reliability in product design had been considered seriously. It will help to understand why the reliability concept of modern product is critical.

- Space Shuttle Challenger: This debacle occurred in 1986, in which all crew members lost their lives. Sadly, many Americans are suffering from low self-esteem because of failure. The main reason for this disaster was design defects of rubber O-rings under cold winter (Table 2.1).

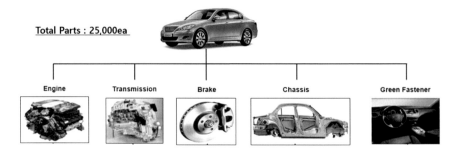

Fig. 2.3 Breakdown of passenger automobile with multi-modules

2.2 Reliability Disasters

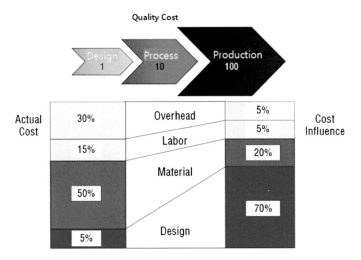

Fig. 2.4 Leverage in product design: total cost of product is determined by its design (approximately 70%)

Table 2.1 Summary of space Shuttle challenger disaster

	Reliability disaster
Phenomenon	
Structure	
Root cause	Failure of two rubber O-rings (Environment conditions: a cold launch day)

Table 2.2 Summary of Chernobyl nuclear reactor explosion

	Reliability disaster
Phenomenon	*(aerial photograph of destroyed reactor with inset map of Europe)*
Structure	*(diagram of reactor showing control rods, radiation shield and containment structure, steam separator, steam, water, graphite moderator, fuel rods, pump)*
Root cause	Reactor explosion (reactor is jumped to around 30,000 MW thermal)

- Chernobyl Nuclear Reactor Explosion: This disaster occurred in 1986, in the former Soviet Union, in which 31 lives were lost. The debacle was the result of design defects such as faulty switch in reactor design (Table 2.2).
- Point Pleasant Bridge Disaster: Bridge located on the West Virginia/Ohio border collapsed in 1967. The disaster resulted in the loss of 46 lives and its basic cause was the metal fatigue of a critical eye bar (Table 2.3).

And in the following sections, we can suggest the numerous other cases for reliability disasters that will help to understand their root causes of modern products.

2.2.1 Versailles Rail Accident in 1842

The Versailles rail accident in 1842 occurred on the railway between Versailles and Paris. Following King Louis Philippe I's celebrations at the Palace of Versailles,

2.2 Reliability Disasters

Table 2.3 Summary of point pleasant bridge disaster

	Reliability disaster
Phenomenon	
Structure	
Root cause	Metal fatigue of a critical eye bar

a train returning to Paris derailed at Meudon. After the leading locomotive broke an axle, the carriages behind piled into it and caught fire. With approximately 200 deaths including that of the explorer Jules Dumont d'Urville, this was the first French railway accident and the deadliest in the world recorded. Since most of passengers wearing the seat also were dead, the accident led to the abandonment of the practice of locking passengers in their carriages. It started the study of metal fatigue subjected to repetitive loads like S–N curve (see Fig. 2.5).

- Root Cause: Metal fatigue of rail was poorly understood at the time and the accident is linked to the beginnings of systematic research into the failure problem.

2.2.2 Tacoma Narrows Bridge in 1940

The Tacoma Narrows Bridge is a pair of twin suspension bridges that connect the city of Tacoma to the Kitsap Peninsula. It went past State Route 16 over the strait and was collapsed by a wind-induced natural frequency on November 7, 1940.

Fig. 2.5 Versailles rail accident (1842) from Wikipedia

The collapse of the bridge had no loss of human life. As recorded on film, this film as a cautionary tale has still been well-known to engineering, architecture, and physics students (see Fig. 2.6).

- Root Cause: without any definitive conclusions, three possible failure causes are assumed

    (1) Aerodynamic instability by self-induced vibrations in the bridge structure
    (2) Periodic eddy formations in bridge
    (3) Random turbulence effects—the random fluctuations by wind velocity of the bridge.

2.2.3 De Havilland DH 106 Comet in 1953

The de Havilland DH 106 Comet was the first commercial jet engine airplane that replaced the propeller plane and could have a transatlantic flight. The Comet prototype first had an aerodynamically design with four turbojet engines in two wings, an aerodynamic fuselage, and large square windows. It first flew on 27 July, 1949. For the 1952 appearance, it offered a quiet and comfortable passenger cabin.

One year later, the Comets began to suffer the design problems that three airplanes were breaking up in flight. Due to airframe metal fatigue, the Comet eventually discover the design flaws at the corners of the square windows subjected

2.2 Reliability Disasters

Fig. 2.6 Tacoma narrows bridge (1940) from Wikipedia

to repetitive stresses. As a result, the Comet was redesigned with oval windows, structural reinforcement, and other changes (see Figs. 2.7 and 2.8).

- Root Cause: fine cracks near the fixed nails of large square windows → repeated pressurization and decompression in airplane → spreading cracks → limit crack → blast in air by the broken window of airplane.

2.2.4 G Company and M Company Rotary Compressor Recall in 1981

In 1981, market share and profits in G company appliance division were falling. For example, making refrigerator compressor required 65 min work of labor in comparison to 25 min for competitors in Japan and Italy. Moreover, labor costs of G Company were higher than that of Japan companies. The alternatives were to purchase compressors from a better design model of Japan or Italy. By 1983, G Company was decided to build a new rotary compressor in-house along with a commitment for a new $120 million factory. G Company and a rival M company had invented the rotary compressor technology that had been using it in air-conditioners for many years.

Fig. 2.7 De Havilland DH 106 Comet (1954) from Wikipedia

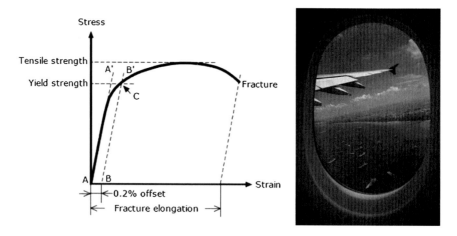

Fig. 2.8 Stress–strain curve and the modified oval window in airplane

A rotary compressor had the less weighted part which was one-third fewer and was more energy efficient than the conventional reciprocating compressors. The rotary compressors took up less space, thus providing more room inside the refrigerator and better meeting customer requirements. The rotary compressor for refrigerator was nearly identical to that used in air conditioners.

However, in a refrigerator, the coolant flows only one-tenth fast and the unit runs about four times longer in one year than an air conditioner. Two small parts inside

2.2 Reliability Disasters

Table 2.4 G company and M company rotary compressor recall summary

	G company	M company
Product	Household refrigerator	
Unit	Rotary comp (sealed refrigerant compressor)	
Production date	1986.3	1985.1
Issued date	1987.7	1991.10
Failed cost	450 million $	560 million $
Failed amount	1.1 million	1 million
Failure mechanism	Abnormal wear out (sintered iron)	Wear out (lubrication at high temp)
	Oil reaction/sludge imbedding	Oil reaction/sludge imbedding
User environment	Worst case	Worst case
After Disaster	Withdraw comp BIZ	Lock out factory

the compressor were made out of powdered metal rather than the hardened steel and cast iron used in air conditioners because this material could be much closer tolerances and reduce the machining costs. The design engineers did not consider the critical failure in early product until the noise claims of domestic house in 1987.

When a rotary compressor was abnormally locking in 1987, G Company and M Company experienced massive recalls of the rotary compressor. As the oil sludge in the refrigeration system blocked the capillary tube, the cooling capacity of the refrigerator decreased. In the compressor development process, reproducing this failure mode and preventing the blocking of this tube were very important to the reliability of the refrigerator. However, reliability testing methods such as the parametric ALT were not used at that time (Table 2.4).

- Root Cause: Abnormal wear out at sintered iron (new parts).

2.2.5 Firestone and Ford Tire in 2000

In the early 2000, the Firestone and Ford tire experienced an unusually tire failures on the Ford Explorer equipped with Firestone tires. The Ford Motor Company had a historically good relationship with Firestone. As Firestone became a subsidiary of Japanese tire manufacturer Bridgestone in 1988, they drifted apart. The U.S. National Highway Traffic Safety Administration (NHTSA) contacted Ford in May 2000 and asked about the high incidence of Firestone tire failure on Ford Explorers model. Immediately, Ford found that it had very high failure rates from 15 in Firestone tires models (see Fig. 2.9).

Firestone recalled the millions of tires including 2.8 million Firestone Wilderness AT tires. A large number of lawsuits have been filed against both Ford and Firestone that there had been over 240 deaths and 3000 catastrophic injuries.

(a) Ford Explorer and Firestone Tires

(b) Firestone fallout

Fig. 2.9 Firestone and ford tire controversy from Wikipedia

The actual accidents come from separating a kind of tire tread when cornering on cloverleaf interchange in high speed.

- Root Cause: Remove air from the tires (Minor design change) → Tire heat up → Damage the tire → Interaction of steel and rubber tire → Tread separation.

2.2.6 Toshiba Satellite Notebook and Battery Overheating Problem in 2007

As approximately 41,000 Toshiba laptops were reported for more than 100 cases of melting laptop and minor injuries, Toshiba had to fell massive recalls in 2007.

Fig. 2.10 Toshiba Satellite T130 notebook and battery overheating problem

The basic cause might overheat and expose a burning to consumers. The root cause came from the heat when processors and batteries run. Laptops are designed to provide adequate airflow for the fan and eliminate the overheating from the case. However, due to the requirements of slim, less weight and compact design, notebooks will push heat-generating components into a smaller space (See Fig. 2.10).

- Root Cause: pushing so much processing power and battery into such a small space (design problem).

2.2.7 Toyota Motor Recalls in 2009

The recalls of automobiles by Toyota Motor Corporation occurred in 2009 for approximately 5.2 million vehicles—the pedal entrapment/floor mat problem, and for 2.3 million vehicles—the accelerator pedal problem. As Toyota widened the recalls to include 1.8 million vehicles in Europe and 75,000 in China, total recall number of cars in the world were considerable 9 million. The U.S. National Highway Traffic Safety Administration (NHTSA) reached to conclusion that pedal misapplication was found responsible for most of the incidents (see Fig. 2.11).

- Root Cause: the pedal entrapment/floor mat problem.

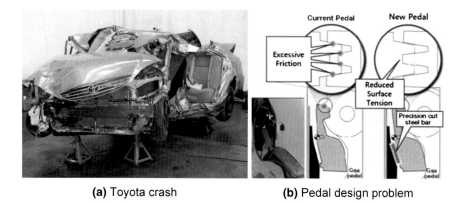

(a) Toyota crash (b) Pedal design problem

Fig. 2.11 Recalls of automobiles by Toyota Motor Corporation

2.3 Development of Reliability Methodologies in History

2.3.1 In the Early of 20s Century—Starting Reliability Studies

The modern concept for reliability was beginning in 1816. The word "reliability" was first coined by poet Samuel Taylor Coleridge [1]. At that time reliability in statistics was defined as the consistency of a set of measurements to express a test. A test is reliable if the same result is repeated. For instance, if a test is designed to measure special marks, the results should be approximately nearly identical to the one. Reliability was a common concept that had been perceived as an attribute of a product. Before taking up the main subject, the milestones of the historical reliability technology in the past century might be briefly summarized in Table 2.5.

In the early times, reliability disasters were the rail accident that France Versailles frequently occurred in 1842. August Wöhler investigated the causes of fracture in railroad axles and started the first systematic studies of S–N Curve (or Wöhler Curve) [2, 3]. To prevent the railroad disasters, S–N curve of materials can be used to minimize the fatigue problem by lowering the stress at critical points in a component. Griffith during World War I developed fracture mechanics to explain the failure of brittle materials. He suggested that the low fracture strength observed in experiments was due to the presence of microscopic flaws in the bulk material that can be still useful (Fig. 2.12) [4].

$$\sigma_f \sqrt{a} \approx C \tag{2.1}$$

where σ_f is the failure stress.

Failure occurs when the free energy attains a peak value at a critical crack length.

2.3 Development of Reliability Methodologies in History

Table 2.5 History summary of reliability technology

~1950	–	Wilhelm Albert publishes the first article on fatigue (1837) A. Wöhler summarized fatigue test results on railroad axles (1870) O.H. Basquin proposes a log-log relationship for S–N curves (1901) John Ambrose Fleming invented vacuum tubes in 1904 Griffith'st theory of fracture (1921) A.M. Miner introduces a linear damage hypothesis (1945)
WW II	Germany	V-I, V-II rocket development (R. Lusser's law)
WW II	US	Reliability of the electron power tube (aircraft electronic devices failure in the WW II)
1954	Japan	Surveys and studies on the electron power tube reliability in the vacuum committee of the institute of electrical engineers
1952–1957	US	US DOD formed the advisory group on the reliability of electronic equipment (AGREE) AGREE suggest vacuum tube follows the bathtub curve
1954	US	First national symposium on reliability and quality control, New York
1950s	US	Several conferences began to focus on various reliability topics (e.g., 1955 Holm conference on electrical contacts)
1961	Italy	The Rome Air Development Center (RADC) introduced a PoF program
1962	US	Launched the Apollo program(FMEA & FTA), first reliability and maintainability conference
1962	US	First symposium on physics of failure in electronics, Chicago
1965	IEC	Reliability and maintainability technical committee, TC 56, Toy
1968	–	Tatsuo endo introduces the rain-flow cycle count algorithm
1971	Japan	First reliability and maintainability symposium

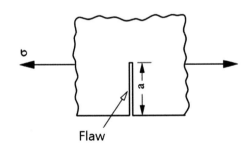

Fig. 2.12 An edge crack (flaw) of length a in a material

$$C = \sqrt{\frac{2E\gamma}{\pi}} \qquad (2.2)$$

where E is the Young's modulus of the material and γ is the surface energy density of the material.

Invented in 1904 by John Ambrose Fleming, vacuum tubes were a basic component for electronics—the diffusion of radio, television, radar, sound reinforcement, sound recording and reproduction, large telephone networks, analog and

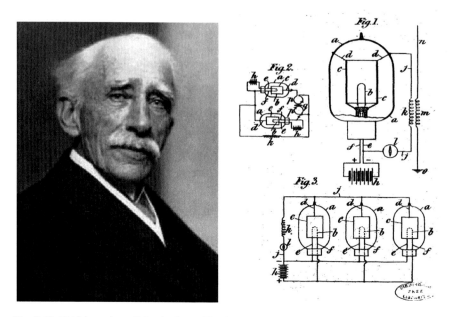

Fig. 2.13 British engineer John Ambrose Fleming and his vacuum tubes patents [5]

digital computers, and industrial process control. The invention of the vacuum tube made modern technologies of product applicable. By 1916, radio with vacuum tubes was used to begin in the public (Fig. 2.13). The concept of reliability by the problematic vacuum tubes began in earnest to develop.

Karl Pearson first mentioned "negative exponential distribution" in 1895. His exponential distribution had a number of interesting properties that were available in the 1950s and 60s. That is, one property of serial system is the ability to add failure rates of different components in product. Simply adding it was rather easily applicable at the time when using mechanical and later electric systems.

$$R(t) = R_1(t) \cdot R_2(t) \cdots R_n(t) \qquad (2.3)$$

$$R(t) = e^{-\lambda_1 t} \cdot e^{-\lambda_2 t} \cdots e^{-\lambda_n t} \qquad (2.4)$$

$$R(t) = e^{-(\lambda_1 + \lambda_2 + \cdots + \lambda_n)t} \qquad (2.5)$$

where R is reliability function, λ is the failure rate, and t is the use time

As automobiles came into more common use in the early 1920s, product improvement by the statistical quality control was introduced by Walter A. Shewhart at Bell Laboratories. He developed the control chart in 1924 and the concept of statistical control. Statistics as a measurement tool would become connected with the development of reliability concepts. While designers were responsible for product quality and reliability, technician took care of the failures.

2.3 Development of Reliability Methodologies in History

Fig. 2.14 A popular automobiles in the early 1920s [6]

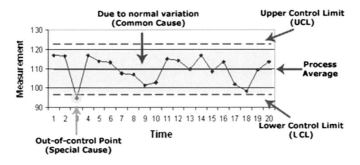

Fig. 2.15 Typical control chart [7]

In the 1930s, quality and process measures in automobile were still growing (Figs. 2.14 and 2.15).

In the 1940s, W. Edwards Deming stressed management responsibility for quality in the military short lecture. He expressed that most of quality problems are actually due to system design errors, not worker error [8]. For instance, an initial reliability concept was applied to the spark transmitters telegraph because of the uncomplicated design. It was a battery powered system with simple transmitters by wire. The main failure mode was a broken wire or insufficient voltage. After WWI, greatly improved transmitters based on vacuum tubes became available (Fig. 2.16).

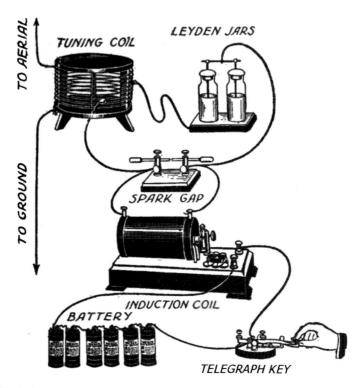

Fig. 2.16 A simple spark-gap transmitter with HV capacitor and output tuning coil

2.3.2 In the World War II—New Electronics Failure in Military

Before World War II, many concepts in reliability engineering still did not exist. However, many new electronic products such as electronic switches, vacuum tube portable radios, radar and electronic detonators are introduced into the military during the WWII (Fig. 2.17).

As the war began, it was discovered that half of the airborne electronics equipment in storage was down in lifetime and unable to meet the military requirements (the Air Core and Navy). Reliability work for this period had to do with new metal materials testing. Study for failure mechanism was only its fatigue or fracture. For instance, M.A. Miner published the seminal paper titled "Cumulative Damage in Fatigue" in 1945 in an ASME Journal. B [9]. Epstein published "Statistical Aspects of Fracture Problems" in the Journal of Applied Physics in February 1948 [10] (Fig. 2.18).

For some more interesting facts, Germany during World War II applied the basic reliability concepts to improve reliability of their V1 and V2 rockets that consist of multi-modules (Fig. 2.19). To complete the mission of V1 and V2 rockets,

2.3 Development of Reliability Methodologies in History

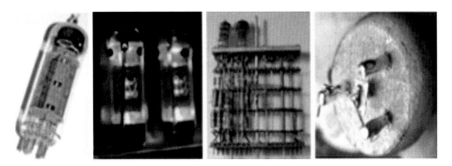

Fig. 2.17 Reliability metric tailored to the leading electronic technology of the world war II and 1950s—vacuum tube and its assembly, discreet transistor & diodes

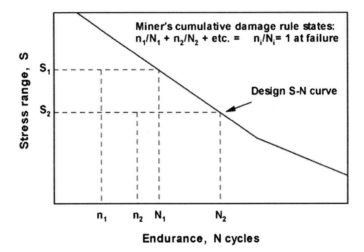

Fig. 2.18 Palmgren-miner linear damage hypothesis

Germany engineer had to improve their reliability. One of reliability theory was Robert Lusser's law. By his law, the reliability of a series system is equal to the product of the reliability of its component subsystems. It means that a series system is "weaker than its weakest link", as the product reliability of a series of components can be less than the lowest value component. After World War II, the United States Department of Defense seriously recognized the necessity for reliability improvement of its military equipment. This law became theoretical basis of MIL-HDBK-217 and MIL-STD-756.

Lusser's Law

Lusser's low (multiplication principle) is a prediction of reliability named after Robert Lusser.
It states that the reliability of a series system is equal to the product of the reliability of its component subsystems, if their failure modes are known to be statistically independent. This method is similar to the

$$R = R1 \times R2 \times R3 \times R4 \times R5 \times \cdots \cdots RX$$

Deterioration makes reliability go down.

Fig. 2.19 V-1 and V-2 missiles and Lusser's law

2.3.3 In the End of World War II and 1950s—Starting the Reliability Engineering

In the start of the 1950s, the main military applications for reliability were the vacuum tube in radar systems or other electronics because these systems proved problematic and costly during the World War II (Fig. 2.20). The vacuum tube computers that had a 1024 bit memory were invented to fill a large room and consume kilowatts of power, though grossly inefficient in modern.

During the war, vacuum tubes mounted in these airplanes had been proved as the problematic parts, though the component price was cheap. After the war, half of the electronic equipment for shipboard failed in lifetime. Failure modes of vacuum tubes in sockets were intermittent working problems. The action plans for a failed electronic system were to explode the system, removing the tubes and reinstalling them at the proper time. In the renovation process, because the military had to consider the cost issues, the operation and logistics costs for the vacuum tubes would become huge. To solve the vacuum tube, Institute of Electrical and Electronic Engineers (IEEE) in 1948 formed the Reliability Society.

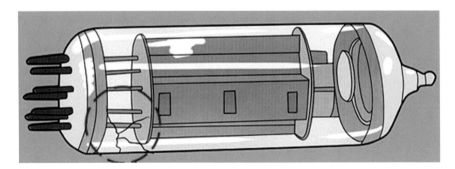

Fig. 2.20 Typical vacuum tube failure—air leakage into the tube due to the crack tube

Z.W. Birnbaum in 1948 had founded the Laboratory of Statistical Research at the University of Washington, which served to use the concept of statistics and probability. In 1951, to study reliability problems with the Air Force, Rome Air Development Center (RADC) was established in Rome and New York.

In 1950, a study group in military was initiated, which was thereafter called the Advisory Group on the Reliability of Electronic Equipment (AGREE). By 1959, reports of this group suggested the following three recommendations for the reliable systems such as vacuum tube: (1) there was a need to develop reliable components for supplier, (2) the military should establish quality and reliability requirements (or specifications) for component suppliers, and (3) actual field data should be collected on components to search out the root causes of problems.

A definition of product lifetime originally came from 1957 AGREE Commission Report. Task Group 1 in AGREE has developed minimum-acceptable measures for the reliability of various types of military electronic equipment, expressed in terms of Mean Time Between Failures (MTBF), though defining lifetime for electronic components were currently inadequate. The final report of AGREE committee suggested the reliability of product as most vacuum tube followed the bathtub curve. Consequently, reliability of components is often defined as "the bathtub curve" that has early failure, useful life, and wearout failure (Fig. 2.21). Today the cumulative distribution function corresponding to a bathtub curve was replaced with a Weibull chart in reliability engineering.

In the early of 1950s, a conference on electrical contacts and connectors was initiated to study the reliability physics—failure mechanisms and reliability topics. In 1955, RADC issued "Reliability Factors for Ground Electronic Equipment." by Joseph Naresky [11]. The conference was publishing proceedings, entitled as "Transaction on Reliability and Quality Control in Electronics", merged with an IEEE Reliability conference and became the Reliability and Maintainability Symposium.

As television in the 1950s was introduced, more vacuum tubes were utilized in America house. Repair problems were often due to the failure of one or more

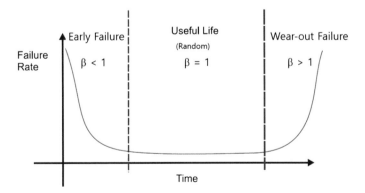

Fig. 2.21 Bathtub curve for vacuum tube radio systems

vacuum tubes. Vacuum tube was a critical switching device that controls electric current through a vacuum in a sealed container—cathode ray tube. Typical reliability problems of the tubes with oxide cathodes evolved as (1) reduce its ability to emit electrons, (2) a stress-related fracture of the tungsten wire, (3) air leakage into the tube, and (4) glowing plate—a sign of an overloaded tube. Most vacuum tube in radio systems followed a bathtub-type curve that was easy to develop replaceable electronic modules—Standard Electronic Modules (SEMs), and then restore a failed system.

AGREE committee also recommended to formally test the products with statistical confidence. And it would carry out the environmental tests that have ultimate temperature and vibration conditions, which became Military Standard 781. The AGREE report originally stated that the definition for reliability is "the probability of a product performing without failure a specified function under given conditions for a specified period of time".

Robert Lusser, Redstone Arsenal, pointed out that 60% of the failures of one Army missile system were due to components that reported on "Predicting Reliability" in 1957. He also stressed that current quality methods for electronic components were inadequate and that new concepts for electric conponents was implemented. ARINC set up an improvement process with vacuum tube suppliers and reduced infant mortality removals by a factor of four. This decade ended with Radio Corporation of America (RCA) publishing information in TR1100 on the failure rates of some military components. RADC used these concepts, which became the basis for Military Handbook 217. Over the next several decades, Birnbaum made suggestions on Chebychev's inequalities, nonparametric statistics, reliability of complex systems, cumulative damage models, competing risk, survival distributions, and mortality rates.

Waloddi Weibull was working in Sweden and investigated the fatigue of materials. He created a Weibull distribution. In 1939, Waloddi Weibull suggested a simple mathematical distribution, which could represent a wide range of failure characteristics by changing two parameters. The Weibull failure distribution does not apply to every failure mechanism but it is useful tool to analyze many of the reliability problems. In 1951, he presented his most famous papers to the American Society of Mechanical Engineers (ASME) on Weibull distribution with seven case studies. Between 1955 and 1963, he investigated the fatigue and creep mechanisms of materials. He derived the Weibull distribution on the basis of the weakest link model of failures in materials. By 1959, he produced "Statistical Evaluation of Data from Fatigue and Creep Rupture Tests: Fundamental Concepts and General Methods" as a Wright Air Development Center Report 59-400 for the US military [12].

In 1961, Weibull published a book on materials and fatigue testing while working as a consultant for the US Air Force Materials Laboratory [13]. The American Society of Mechanical Engineers awarded Weibull their gold medal in 1972. The Great Gold Medal from the Royal Swedish Academy of Engineering Sciences was personally presented to him by King Carl XVI Gustaf of Sweden in 1978 (Fig. 2.22).

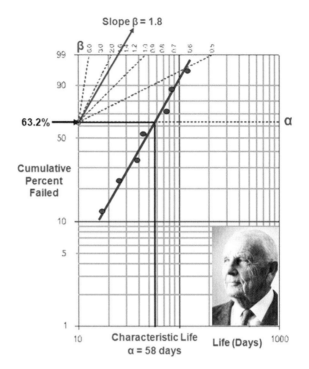

Fig. 2.22 Sweden engineer Waloddi Weibull and his Weibull distribution

As Weibull analysis methods and applications were propagating, a number of people began to use the Weibull chart. Dorian Shain wrote an early booklet on Weibull in the late of 1950s, while Leonard Johnson at General Motors helped improve the plotting methods by suggesting median ranks and beta Binomial confidence bounds in 1964. Dr. Robert Abernethy developed a number of applications, analysis methods, and corrections for the Weibull function. Professor Gumbel demonstrated that the Weibull distribution is a Type III Smallest Extreme Value distribution such as Eqs. (2.6) and (2.7) [14]. Dr. Robert Abernethy developed a number of applications, analysis methods and corrections for the Weibull function

$$F(x) = \exp\left(-\left(\frac{a-x}{b}\right)^c\right) \tag{2.6}$$

where $x \leq a, b > 0, c > 0$

$$K = A \exp\left(-\frac{E_a}{RT}\right) \tag{2.7}$$

where k is the rate constant of a chemical reaction, T is the absolute temperature, A is the pre-factor, E_a is the activation energy, and R is the universal gas constant.

In 1963, Weibull was a visiting professor at Columbia and there worked with professors Gumbel and Freudenthal in the Institute for the Study of Fatigue and Reliability. While he was a consultant for the US Air Force Materials Laboratory, he published a book on materials and fatigue testing and the related reports till 1970.

Aeronautical Radio, Incorporated (ARINC) set up an improvement process with vacuum tube suppliers to reduce its infant mortality removals. As publishing information in TR1100 on the failure rates of some military components, it became the basis for Military Handbook 217—"Reliability Prediction of Electronic Equipment." Navy Military published Handbook 217 in 1962. Papers for electronic components were being published at conferences: "Reliability Handbook for Design Engineers" published in Electronic Engineers, in 1958 by F.E. Dreste and "A Systems Approach to Electronic Reliability" by W.F. Leubbert in the Proceedings of the IRE (1956) [15]. C.M. Ryerson produced a history of reliability to 1959 in the proceedings of the IRE entitled as Proceedings of the IEEE [16].

2.3.4 In the 1960s and Present: Mature of Reliability Methodology—Physics of Failure (PoF)

Physics of Failure (PoF) for electronic components in 1960s started with several significant events—invention of the transistor in 1947 and transistor radio in 1954, which became the most popular electronic communication device during the 1960s and 1970s. People with pocket size transistors listened to music everywhere (Fig. 2.23). These devices had some problems—electromechanical faults, transistor failure, and capacitor problems. POF is a kind of systematic approach to the design and development of reliable product to prevent failure. Based on the knowledge for the root cause of failure mechanisms, electronic system can improve its preformance.

RADC worked in earnest the Physics of Failure in Electronics Conference sponsored by Illinois Institute of Technology (IIT). In the 1960s America strong commitment to space exploration would turn into National Aeronautics and Space Administration (NASA), a key efforts to improve the reliability of components and systems that could work properly to complete the space missions. RADC produced the document "Quality and Reliability Assurance Procedures for Monolithic Microcircuits." Semiconductors were a popular use in small portable transistor radios. Next, low-cost germanium and silicon diodes were able to meet the requirements. Dr. Frank M Gryna published a Reliability Training Text through the Institute of Radio Engineers (IRE).

In this period, the nuclear power industry and the military—missiles, airplanes, helicopters and submarine applications enabled the reliability problems of a variety of technologies to initiate POF. The study of EMC (Electro-Magnetic Compatibility) system effects was initiated at RADC in the 1960s (Fig. 2.24).

One of the milestones in the Proceedings of the 7th National Symposium of Reliability and Quality Control was the proof of the effectiveness of the Arrhenius

2.3 Development of Reliability Methodologies in History

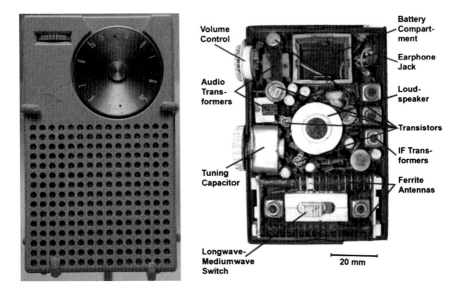

Fig. 2.23 A transistor radio with multiple parts from Wikipedia

model for semiconductors in 1962. G.A. Dodson and B.T. Howard of Bell Labs published the papers, entitled as "High Stress Aging to Failure of Semiconductor Devices." [17] This conference issued lots of other papers. It could look at the technical improvement of other electronic components, and renamed as the Reliability Physics Symposium (RPS) in 1967. Shurtleff and Workman in the late 1960s issued the original paper on step stress testing that establishes limits when applied to Integrated Circuits.

Fig. 2.24 Andy Grove, Bruce Deal, and Ed Snow at the Fairchild Palo Alto R&D laboratory and first commercial metal oxide semiconductor (MOS) IC in 1964 from Wikipedia

Electro-migration in electronic system is one of failure mechanism, which applied to the transport of mass in metals when the metals are stressed at high current densities. J.R. Black published his work on the physics of electro-migration in 1967. Since the number of free charge carriers increases with temperature, silicon in semiconductor began to dominate reliability activities for a variety of industries. The U.S. Army Material Command issued a Reliability Handbook (AMCP 702-3) in 1968. On the other hands, Shooman's Probabilistic Reliability also was issued to explain statistical methods.

To investigate the failure mode of electronic components, automotive industry published a FMEA handbook for technical improvement of suppliers, not yet published as a Military standard. As a series of commercial satellites were launched, the reliability study for communications was strengthened by International Telecommunications Satellite Organization (INTELSAT) that was providing international broadcast services between the U.S. and Europe in 1965. Professionals around the world took part in reliability conferences. As Apollo was landing a moon, people recognized how far reliability had progressed in the recent decade.

$$f(t) = \frac{1}{2\mu^2\gamma^2\sqrt{\pi}} \left(\frac{t^2 - \mu^2}{\sqrt{\frac{t}{\mu}} - \sqrt{\frac{\mu}{t}}} \right) \exp\left[-\frac{1}{\gamma^2} \left(\frac{t}{\mu} + \frac{\mu}{t} - 2 \right) \right] \qquad (2.8)$$

where γ is a shape parameter, μ is a scale parameter

As seen in Eq. (2.8), in 1969, Birnbaum and Saunders suggested a life distribution model that could be derived from a physical fatigue process where crack growth causes failure. Since one of the best ways to choose a life distribution model is to derive it from a physical/statistical argument that is consistent with the failure mechanism, the Birnbaum-Saunders fatigue life distribution is worth considering.

As the microcomputer had been invented in the 1970s, RAM memory size was growing at a rapid rate. Vacuum tube was replaced with Integrated Circuit (IC). The variety of ICs—Bipolar, NMOS, and CMOS increased very rapidly. In the middle of the 1970s, Electrostatic discharge (ESD) and Electrical Over Stress (EOS) were discussed by some papers and eventually became the hot issues of a conference in the decade end.

In the same manner, studies for passive components—resistor, inductor, and capacitor in International Reliability Physics Symposium (IRPS) moved to a Capacitor and Resistor Technology Symposium (CARTS). The progressive papers on gold aluminum inter-metallics, accelerated testing, and the use of Scanning Electron Microscopes (SEM) were in a few highlights of the decade.

In middle of 1970s, Hakim and Reich published a paper on the evaluation of plastic encapsulated transistors and ICs on field data. And two most memorable reliability papers were one on soft-errors from alpha particles first reported by Woods and May and on accelerated testing of ICs with activation energies calculated for a variety of failure mechanisms by D.S. Peck. In the end of the decade, Bellcore collected commercial field data and became the basis of the Bellcore reliability prediction methodology used widely with MIL-STD-217F.

2.3 Development of Reliability Methodologies in History

During the Apollo space program, the spacecraft and its components worked reliably all the way to the moon and back. In coming to the Navy, all contracts should contain specifications for reliability and maintainability instead of just performance requirements. Military Standard 1629 on FMEA was issued in 1974, NASA made great strides at designing and developing spacecraft such as the space shuttle. Their emphasis was on risk management through the use of statistics, reliability, maintainability, system safety, quality assurance, human factors, and software assurance. Reliability had expanded into a number of new areas as technology rapidly advanced. Emphasizing temperature cycling and random vibration became ESS testing, eventually issued as a Navy document P-9492 in 1979 and make a book on Random Vibration with Tustin in 1984. The older quality procedures were replaced with the Navy Best Manufacturing Practice program.

In the 1980s, televisions had become product that used all semiconductors. Automobiles rapidly increased their use of semiconductors with a variety of microcomputers. Large air conditioning systems, microwave ovens, and a variety of other appliances developed one chip electronic controllers. Communications systems began to adopt electronics to replace older mechanical switching systems. Bellcore issued the first consumer prediction methodology for telecommunications and SAE developed a similar document SAE870050 for automotive applications.

As seen in Fig. 2.25, during this decade, as the failure rate of many electronic components including mechanical components were dropped by a factor of 10, engineer questioned on the bathtub curve. For such a situation, the traditional failure rate typified by the bathtub curve can be reduced to resemble the failure rate represented by a flat, straight line with the shape parameter β.

Software became important to the systems improvement by advancing with work at RADC. Software reliability developed models such as Musa Basic to predict the number of missed software faults that might remain in code. The Naval Surface Warfare Center issued Statistical Modeling and Estimation of Reliability Functions for Software in 1983. Contributions by William Meeker, Gerald Hahn, Richard Barlow, and Frank Proschan developed models for wear, degradation, and system reliability.

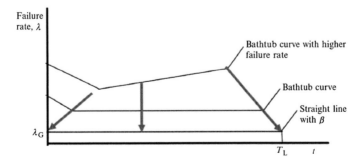

Fig. 2.25 Bathtub curve and straight line with slope β

The PC came into dominance as a tool for measurement & control and enhanced the possibility of canned programs for evaluating reliability. In the end of the decade, FMEAs, FTAs, reliability predictions, block diagrams, and Weibull Analysis programs were performed in the commercial use. The challenger disaster caused people to recognize the assessment significance of system reliability. Many of the military specifications–Military Handbook 217 became obsolete and best commercial practices were often adopted. Most industries developed their own reliability standards like the JEDEC Standards for semiconductors and the Automotive Standard Q100 and Q101. Afterward, in the last century, the rise of the Internet created a variety of new challenges—micro-electro mechanical systems (MEMS), handheld GPS, and handheld devices—for reliability. Consumers have become more aware of reliability disasters. In many ways, reliability became part of everyday life and consumer expectations. The developed methodology in reliability engineering has widely been still developing until now. However, new methodology for reliability is still required to find the problematic parts that is the main cause of recall before production.

References

1. Saleh JH, Marais K (2006) Highlights from the early (and pre-) history of reliability. Eng Reliab Eng Syst Saf 91(2):249–256
2. Wöhler A (1855) Theorie rechteckiger eiserner Brückenbalken mit Gitterwänden und mit Blechwänden. Z Bauwsn 5:121–166
3. Wöhler A (1870) Über die Festigkeitsversuche mit Eisen und Stahl. Z Bauwsn 20:73–106
4. Griffith AA (1921) The phenomena of rupture and flow in solids. Philos Trans R Soc Lond Ser A 221:163–198
5. Fleming, JA U.S. Patent 803,684, 17 Nov 1815
6. Ford (1929) 1930 model brochure—beauty of line—mechanical excellence. Retrieved 24 May 2012
7. Shewhart WA (1931) Economic control of quality of manufactured product. D. Van Nostrand Company, New York
8. Deming WE, Stephan F (1940) On a least squares adjustment of a sampled frequency table when the expected marginal totals are known. Ann Math Stat 11(4):427–444
9. Miner MA (1945) Cumulative damage in fatigue. J Appl Mech 12(3):59–64
10. Epstein B (1948) Statistical aspects of fracture problems. J Appl Phys 19(2):140–147
11. Naresky JJ (1962) Foreword. In: Proceedings of first annual symposium on the physics of failure in electronics, 26–27 Sept
12. Weibull W (1959) Statistical evaluation of data from fatigue and creep rupture tests, part I: fundamental concepts and general methods. Wright air development center technical report 59–400, Sweden, September
13. Weibull W (1961) Fatigue testing and analysis of results. Pergamon Press, London
14. Abernethy R (2002) The new Weibull handbook. 4th edn self published. ISBN 0-9653062-1-6
15. Lloyd D, Lipow M (1962) Reliability: management, methods and mathematics. Prentice Hall, Englewood Cliffs
16. Knight R (1991) Four decades of reliability progress. In: Proceedings of annual RAMS, pp 156–160
17. George E (1998) Reliability physics in electronics: a historical view. IEEE Trans Reliab 47(3):379–389

Chapter 3
Modern Definitions in Reliability Engineering

Abstract This chapter will briefly review the modern definitions in reliability engineering that can be used widely—bathtub, Mean Time Between Failure (MTBF), fundamentals in statistics and probability theory, statistical distributions like Weibull, and time-to-failure model. From customer's standpoint, when a product is delivered to the end-user, reliability explained as lifetime and failure rate can be assessed through product specification. Many reliability concepts used in the product predict the failure rate/lifetime of components subjected to random loads. So the product reliability is related to the robust design of mechanical system without design defects in lifetime. For mechanical engineer, reliability theory may feel complex because of concepts of probability and statistics. However, the reliability concepts are required to develop new methodology of reliability assessment in product. It will help to establish the testing method of reliability that points out the design failure or reliability disasters in the reliability-embedded developing process. As a design parameter method, Taguchi method has developed many concepts but it still has the weak points. Based on Taguchi concepts, new reliability methodology is still required to discover the design problem of parts.

Keywords Reliability concepts · Reliability engineering · Probability theory · Robust design

3.1 Introduction

Reliability is the ability of an item to work properly the intended functions during its lifetime. The modern concepts in reliability engineering started through the reliability study of vacuum tube in the WW2. The reliability concepts except the quality control in product manufacture can focus on the study of product quality itself in design.

3.1.1 Bathtub Curve

To describe three types of the failure rate (or hazard function in product, the bathtub curve in reliability engineering comprises as (Fig. 3.1):

- Early failures in the first part are a decreasing failure rate ($\beta < 1$).
- Random failures in the second part are a constant failure rate ($\beta = 1$).
- Wear out failures in the third part are an increasing failure rate ($\beta > 1$).

As the failure of the vacuum tube was studied in WWII, the bathtub curve was created by mapping the rate of early "infant mortality" failures, the rate of random failures with constant failure rate during its "useful life," and finally the rate of "wear out" failures as the product exceeds its design lifetime. In the early life of a product following to the bathtub curve, the failure rate is high but rapidly decreasing as defective products are removed, and early sources of potential failure such as storage, handling and installation error are dominated. In the midlife of a product—generally, once it reaches consumers—the failure rate is low and constant. In this period product may experience the catastropic disaster if design problem in product exist. Before production, design problem should be found by proper method-accelerated life testing. Product lifetime will increase. In the late life of the product, the failure rate increases. Thus, there are three types of reliability testing in accordance with the failure rate (Fig. 3.2).

- Early failures: Because it requires the short test time, it easily improve prior to shipment

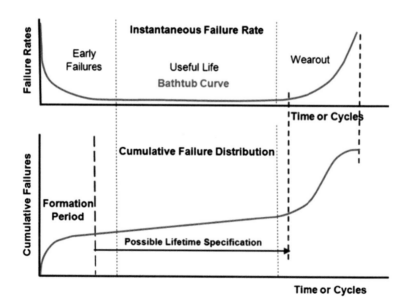

Fig. 3.1 Bathtub curve

3.1 Introduction

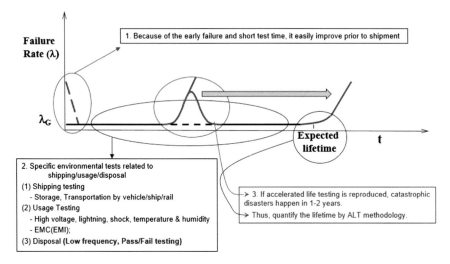

Fig. 3.2 Three types of reliability tests in accordance with the failure rate

- Random failures: Specific environmental tests related to shipping/usage/disposal
  - Shipping testing: storage, transportation by vehicle/ship/rail
  - Usage Testing: High voltage, lightning, shock, temperature and humidity, EMC (EMI)
  - Disposal (Low frequency, Pass/Fail testing)
- Catastrophic disasters: Catastrophic disasters often happen in 1–2 years. It comes from the design missing. If it is reproduced by accelerated life testing described in Chap. 7, it can be eliminated.

The term "Military Specification" is also used to describe systems in which the infant mortality section of the bathtub curve has been burned out or removed. This is done mainly for life critical or system critical applications as it greatly reduces the possibility of the system failing early in its life. Manufacturers will do this at some cost generally by means similar to environmental stress screening. In reliability engineering, the cumulative distribution function (CDF) corresponding to a bathtub curve may be transformed and analyzed by using a Weibull chart.

3.2 Fundamentals in Probability Theory

The failure behavior of product and components can be expressed as the statistical and probability theory because of these random events in field. In market data it is proper to know the failure characteristics of sample data from a population of items. For example, if hundred televisions put on test and 12 among them fail, analyze the times to failure. If thousand aircraft engine controllers are operating in service, collect all the times to failure data and analyze them. Here test data may be not only times but distance or cycles, etc.

3.2.1 Probability

The probability was originally established by gamblers who were interested in high stakes. To answer the question "how probable," it is that a certain event A occurs in a game of gambling. An early mathematician, Laplace and Pascal, invented the probability. That is, when N is the number of times that X occurs in the n repeated experiments, the probability of occurrence of event X, $P(X)$, can be defined as:

$$P(X) = \lim_{n \to \infty} \left(\frac{\text{number of cases favorable to } X}{\text{number of all possible cases}} \right) = \lim_{n \to \infty} (N/n), \quad (3.1)$$

where X is a random variable.

For example, if trials n approaches ∞, the probability of rolling a 1 with a die is:

$$P(X = 3) = \frac{1}{6} = 0.167 \quad (3.2)$$

In general, this is adequate for gambling. On the other hands, in technical reality, the failure probabilities happen to vary amounts. In modern theory, probability is seen as a basic principle and has the assumptions:

- Each random variable X has $0 \leq P(X) \leq 1$.
- The area under the curve is equal to 1: $\int P(X) dX = 1$, where $0 \leq X \leq \infty$
- If $X_1, X_2, X_3, \ldots$ are random variables, then $P(X_1 \cup X_2 \cup \cdots) = P(X_1) + P(X_2) + \cdots$
- Area under the curve between two values is the probability: $P(a \leq X \leq b) = \int_b^a f(X) dX$

If not all data is normally distributed, other distributions-Weibull analysis is especially suited to failure rates. Select the failure data and draw histogram. We can find the skewed right (or left) histogram like Weibull distribution. When failure behavior is represented graphically, basic probability concepts are mean, median, mode, standard deviation (see Fig. 3.3).

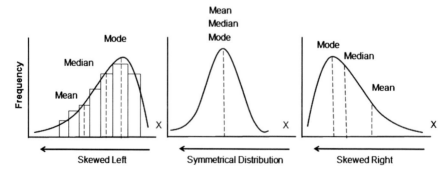

Fig. 3.3 Mean, median, and mode for skewed left/right and symmetric distribution

3.2 Fundamentals in Probability Theory

3.2.1.1 Mean

For a data set, mean refers to one measure of the central tendency either of a probability distribution or of the random variable characterized by that distribution. If we have a data set containing the failure times $t_1, t_2, \ldots, t_n$, the mean is defined by the formula:

$$t_m = \frac{t_1 + t_2 + \cdots + t_n}{n} = \frac{\sum_{i=1}^{n} t_i}{n} \quad (3.3)$$

The mean describes the parameter where the middle of the failure times approximately locates. The mathematical mean is affected to the lowest or highest failure times.

3.2.1.2 Median

Median is the number separating the higher half of a data sample. In reliability testing, the median is the time in the middle of failure data. The median may be determined by the CDF $F(t)$.

$$F(t_{median}) = 0.5 \quad (3.4)$$

The mathematical median is not affected to the lowest or highest failure times.

3.2.1.3 Mode

The mode is the value that appears most often in a set of data. In reliability testing, the mode is the most frequent failure time. The mode is the maximum value of the density function $f(t)$. So it can be expressed as:

$$f'(t_{mode}) = 0 \quad (3.5)$$

The mathematical median is not affected to the lowest or highest failure times.

3.2.1.4 Standard Deviation

In statistics, the standard deviation (SD) is used to quantify the variation amount of a set of data values. In reliability testing, the standard deviation is the square root of the variance. This is expressed by

$$\sigma = \left[\frac{\sum_{i=1}^{n} (t_i - t_m)^2}{n} \right]^{1/2} \qquad (3.6)$$

The standard deviation has the same dimension as the failure times t_i.

3.2.1.5 Expected Value

In probability theory, the expected value of a random variable is intuitively the average value of the long-run repetition experiment. The expected value, $E(t)$, of a continuous random variable is expressed by

$$E(t) = M = \int_{-\infty}^{\infty} tf(t)\,dt \qquad (3.7)$$

3.2.2 Probability Distributions

Probability distributions are typically defined in terms of the probability density function (PDF). However, there are a number of probability functions used in applications.

3.2.2.1 Reliability Function

The common used function in reliability engineering is the reliability function. This function is the probability of an item operating for a period of time without failure. The reliability function can be expressed as:

$$R(t) = P(T > t) = 1 - F(t) = \int_{t}^{\infty} f(x)\,dx \qquad (3.8)$$

$R(t)$ is the probability that the item will not fail in the interval $(0, t]$. $R(t)$ is the probability that it will survive at least until time t—it is sometimes called the survival function (see Fig. 3.4).

Fig. 3.4 Cumulative distribution function $F(t)$ and reliability function $f(t)$

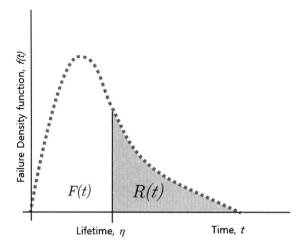

3.2.2.2 Cumulative Distribution Function

The cumulative distribution function (CDF) is the probability that the variable t takes a value less than or equal to T. CDF associated with the time to failure T is expressed as:

$$F(t) = P(T \leq t) \quad (3.9)$$

which is the probability that the system fails within the time interval $(0; t]$. If T is a continuous random variable, the probability function is related to its PDF $f(t)$ by

$$F(t) = \int_0^t f(x)\,dx \quad (3.10)$$

3.2.2.3 Probability Density Function (PDF)

In probability theory, a probability density function (PDF) is a function that describes the relative likelihood for this random variable to take on a given value (Fig. 3.5). In reliability testing, density function $f(t)$ is defined by:

$$\frac{dF(t)}{dt} = \frac{d\int_0^t f(x)\,dx}{dt} = f(t) \quad (3.11)$$

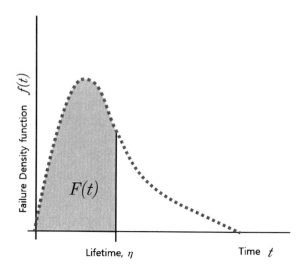

Fig. 3.5 Density function *f*(*t*) and cumulative distribution function *F*(*t*)

3.2.2.4 Failure Rate

Failure rate (or Hazard rate function) is the frequency with which an engineered system or component fails. Consider the conditional probability:

$$P(t < T < t + \Delta t | T > t) = \frac{P(t < T \leq t + \Delta t)}{R(t)} = \frac{F(t + \Delta t) - F(t)}{R(t)} \quad (3.12)$$

In reliability engineering, failure rate (or hazard rate function) $\lambda(t)$ is defined by:

$$\lambda(t) = \lim_{\Delta t \to 0} \frac{P(t < T < t + \Delta t | T > t)}{\Delta t} = \frac{f(t)}{R(t)} \quad (3.13)$$

$\lambda(t)\,dt$ is the probability that the system will fail during the period $(t; t + dt]$, given that it has survived until time t.

3.2.2.5 Cumulative Hazard Rate Function

A survival and hazard function is to analyze the expected duration of time until one or more events happen, such as failure in mechanical systems. Cumulative hazard rate function $\Lambda(t)$ is defined by:

$$\Lambda(t) = \int_0^t \lambda(x)\,dx \quad (3.14)$$

3.2 Fundamentals in Probability Theory

Suppose the failure rate $\lambda(t)$ is known. Then it is possible to obtain $f(t)$, $F(t)$, and $R(t)$.

$$f(t) = \frac{dF(t)}{dt} = -\frac{dR(t)}{dt} \Rightarrow \lambda(t) = -\frac{dR/dt}{R} \qquad (3.15)$$

$$\frac{dR}{R} = -\lambda(t)dt \qquad (3.16)$$

If Eq. (3.16) is integrated, then reliability function becomes

$$R(t) = \exp\left[-\int_0^t \lambda(\tau)d\tau\right] \qquad (3.17)$$

So the density function and CDF are defined as:

$$f(t) = \lambda(t)\exp\left[-\int_0^t \lambda(\tau)d\tau\right] \qquad (3.18)$$

$$F(t) = 1 - \exp\left[-\int_0^t \lambda(\tau)d\tau\right] \qquad (3.19)$$

Relationship between reliability function $R(t)$ and CDF $F(t)$ can be summarized in Fig. 3.6.

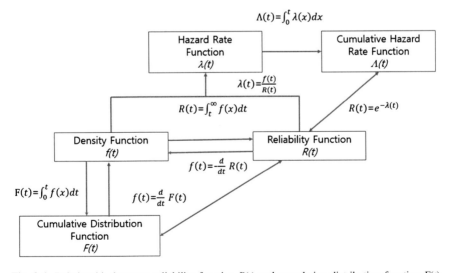

Fig. 3.6 Relationship between reliability function $R(t)$ and cumulative distribution function $F(t)$

3.3 Reliability Lifetime Metrics

An important goal for reliability designers is to assess lifetime against product failures. Reliability lifetime metrics are used to quantify a failure rate and the resulting time of expected performance. MTTF, MTBF, MTTR, FIT, and BX% life are reliability lifetime metrics as follows:

- MTTF (Mean Time To Failure),
- MTBF (Mean Time Between Failure),
- MTTR (Mean Time To Repair),
- BX% life.

3.3.1 Mean Time to Failure (MTTF)

MTTF is a basic lifetime metric of reliability to specify the lifetime of non-repairable systems—"one-shot" devices like light bulbs. It is the mean time until a piece of equipment fails at first statistically. MTTF is the mean over a long period of time with a large unit (Fig. 3.7).

$$\text{MTTF} = \frac{t_1 + t_2 + \cdots + t_n}{n} \quad (3.20)$$

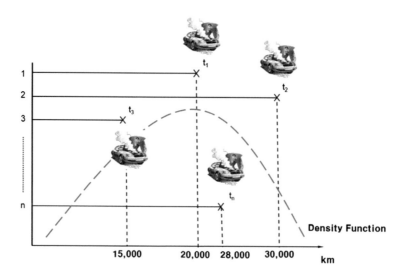

Fig. 3.7 Concept of mean time to failure

3.3 Reliability Lifetime Metrics

As seen in Fig. 3.7, we know that MTTF is 23,000 km if using Eq. (3.20). And the MTTF can be described with other mathematical terms:

$$\text{MTTF} = E(T) = \int_0^\infty t \cdot f(t) dt = -\int_0^\infty t \frac{dR(t)}{dt} dt = \int_0^\infty R(t) dt, \quad (3.21)$$

where $f(t) = \frac{d}{dt} F(t) = -\frac{d}{dt} R(t)$

Example 3.1. Consider a system with reliability function

$$R(t) = \frac{1}{(0.2t+1)^2}, \quad \text{for } t > 0 \quad (3.22)$$

Find the PDF, failure rate, and MTTF

$$\text{Probability ensity } f(t) = -\frac{d}{dt} R(t) = \frac{0.4}{(0.2t+1)^3}$$

$$\text{Failure rate } \lambda(t) = \frac{f(t)}{R(t)} = \frac{0.4}{(0.2t+1)}$$

$$\text{Mean time to failure MTTF} = \int_0^\infty R(t) dt = 5 \text{ months}$$

3.3.2 Mean Time Between Failure (MTBF)

Mean Time Between Failure (MTBF) is a reliability metric used to describe the mean lifetime of repairable components—computers, automobiles, and airplanes. MTBF remains a basic measure of a systems' reliability for most products, though it still is debated and changed. MTBF still is more important for industries and integrators than for consumers (Fig. 3.8).

$$\text{MTBF} = \frac{T}{n} \quad (3.23)$$

MTBF value is equivalent to the expected number of operating hours (service life) before a product fails. There are several variables that can impact failures. Aside from component failures, customer use/installation can also result in failure. MTBF is often calculated based on an algorithm that factors in all of a product's components to reach the sum life cycle in hours. MTBF is considered a system failure. It is still regarded as a useful tool when considering the purchase and

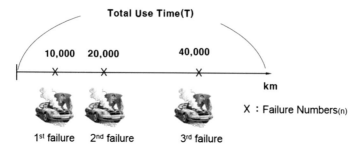

Fig. 3.8 Concept of mean time between failure

installation of a product. For repairable complex systems, failures are considered to be those out of design conditions which place the system out of service and into a state for repair. Technically, MTBF is used only in reference to a repairable item and non-repairable items, while MTTF is used for non-repairable items like electric components.

3.3.3 Mean Time to Repair (MTTR)

Mean Time To Repair (MTTR) is the average lifetime needed to fix a problem. In an operational system, repair generally means replacing a failed hardware part. Thus, hardware MTTR could be viewed as mean time to replace a failed hardware module. Taking too long to repair a product drives up the cost of the installation in the long run, due to down time until the new part arrives and the possible window of time required scheduling the installation. To avoid MTTR, many companies purchase spare products so that a replacement can be installed quickly. Generally, however, customers will inquire about the turnaround time of repairing a product, and indirectly, that can fall into the MTTR category. And relationship among MTTF, MTBF and MTTR can be described in Fig. 3.9.

3.3.4 BX% Life

The BX life metric originated in the ball and roller bearing industry, but has become a product lifetime metric used across a variety of industries today. It is particularly useful in establishing warranty periods for a product. The BX% life is the lifetime metric which takes to fail $X\%$ of the units in a population. For example, if an item has a B10 life of 1000 km, then 10% of the population will have failed by 1000 km of operation.

Alternatively, the B10% life has the 90% reliability of a population at a specific point in product lifetime. The "BX" or "Bearing Life" nomenclature refers to the

3.3 Reliability Lifetime Metrics

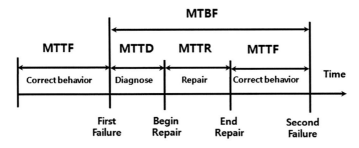

Fig. 3.9 A schematic diagram of MTTF, MTTR, and MTBF

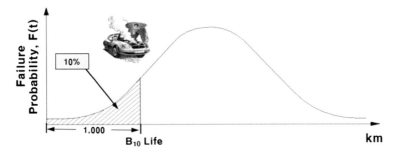

Fig. 3.10 Concept of *BX*% life

time at which *X*% of items in a population will fail. The B10 life metric became popular among product industries due to the industry's strict requirement. Now B1, B10 and B50 lifetime values serve as a measurement for the reliability of a product (Fig. 3.10).

3.3.5 The Inadequacy of the MTTF (or MTBF) and the Alternative Metric BX Life

Two representative metrics of reliability may describe product lifetime and the failure rate. The failure rate are adequate for understanding situations that include unit periods, such as the annual failure rate. But the lifetime is frequently indexed using the mean time to failure.

The MTTF are misinterpreted. For instance, assume that the MTTF of a printed circuit assembly for television is 40,000 h. Annual usage reaches 40,000 divided by 2000 and become 20 years, which is regarded as the lifetime of the unit. The average lifetime of the television PCA is assumed to be 20 years. But because actual customer experience is that the lifetime of a television is a 10 years, this can lead to misjudgments or overdesign that wastes material.

MTTF is often assumed to be the same as lifetime because customers understand the MTTF as, literally, the average lifetime because customers understand the average lifetime of their appliances, so they suppose products will operate well with until they reach the MTTF. In reality, this does not happen. By definition, the MTTF is an arithmetic mean; specifically, it equals the period from the start of usage to the time that the 63rd item fails among 100 sets of one production lot when arranged in the sequence of failure times.

Under this definition, the number of failed televisions before the MTTF is reached would be so high that customers would never accept the MTTF as a lifetime index in the current competitive market. The products of first-class companies have fewer failures in a lifetime than would occur at the MTTF. In the case of home appliances, customers expect no failure for 10 years. The failure of the TV is accepted from the customer's perspective in the later time. Customers would expect the failure of all televisions once the expected use time is exceeded—12 years in the case of a television set—but they will not accept major problems within the first 10 years.

The MTTF is inappropriate as a lifetime index. Alternatively, it is reasonable to define the lifetime as the point in time when the accumulated failure rate has reached $X\%$. This is called the BX life. The value X may vary from product to product, but for home appliance, the time to achieve a 10–30% cumulative failure rate failure rate, B20–30 life, exceeds 10 years. Thus, an average annual failure rate equals to 1–3%.

Now let us calculate the B10 life from the MTTF of 40,000 h. Since the annual usage is 2000 h, the B10 life is 2 years, which means that the yearly failure rate would be 5%. The reliability level of this television, then, would not be acceptable in light of the current annual failure rate of 1–3%. The misinterpretation of reliability using an MTTF of 20 years would lead to higher service expenses if the product were released into the market without further improvement. The lifetime of a television is 12–14 years, not 20 years. Since random failure cannot account for the sharply increasing failure rate, the MTTF based on random failure or on an exponential distribution is obviously not the same as the design lifetime of product (Table 3.1).

Table 3.1 Results of 1987 army SINCGARS study[a]

Vendor	MIL—HDBK-217 MTBF (h)	Actual test MTBF (h)
A	2840	1160
B	1269	74
C	2000	624
D	1845	2174
E	2000	51
F	2304	6903
G	3080	3612
H	811	98
I	2450	472

[a]The transition from Stadterman et al. [1]

3.4 Statistical Distributions

3.4.1 Poisson Distributions

The Poisson distribution is named after Simeon Poisson (1781–1840), a French mathematician, and used in situations where big declines in a time period occurs with a specific average rate, regardless of the time that has elapsed. More specifically, this distribution is used when the number of possible events is large, but the occurrence probability over a specified time period is small. Two examples of such a situation are as follows (Fig. 3.11):

- A store that rents books has an average rental of 200 books every Saturday night. Using this data, you can predict the probability that more books will sell (perhaps 300 or 400) on the following Saturday nights.
- Another example is the number of diners in a certain restaurant every day. If the average number of diners for seven days is 500, you can predict the probability of a certain day having more customers.

A Poisson distribution has the following properties:

- The experiment results in outcomes that can be classified as successes or failures.
- The average number of successes (μ) that occurs in a specified region is known.
- The probability that a success will occur is proportional to the size of the region.
- The probability that a success will occur in an extremely small region is virtually zero.

This distribution also has applications in many reliability areas when one is interested in the occurrence of a number of events that are of the same type. Each

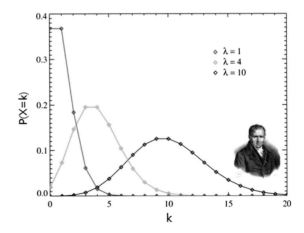

Fig. 3.11 Poisson distributions

event's occurrence is denoted as a timescale and each event represents a failure. The Poisson density function is expressed by

$$f(x) = \frac{(\lambda t)^x e^{-\lambda t}}{x!} \quad \text{for } x = 0, 1, 2, \ldots, n, \tag{3.24}$$

where λ is the constant failure rate, t is the time.

If exponential distribution follows, we can let $m = \lambda t$.

$$P(x, m) = \frac{(m)^x e^{-m}}{x!} \tag{3.25}$$

The CDF is given by

$$F(y) = \sum_{i=0}^{y} \left[(\lambda t)^i e^{-\lambda t} \right] / i! \tag{3.26}$$

In a certain region, the number of traffic accidents averages one per 2 days happens. Find the probability that $x = 0, 1, 2$ accidents will occur in a given day. So the number of traffic accidents averages one per two days, $m = \lambda t = 0.5$,

$$X = 0, \, f(0) = \frac{(0.5)^0 e^{-0.5}}{0!} = 0.606, \tag{3.27}$$

Accident days = 365 day × 0.606 = 221 day

$$X = 1, \, f(1) = \frac{(0.5)^1 e^{-0.5}}{1!} = 0.303, \tag{3.28}$$

Accident days = 365 day × 0.303 = 110 day

$$X = 2, \, f(2) = \frac{(0.5)^2 e^{-0.5}}{2!} = 0.076, \tag{3.29}$$

Accident days = 365 day × 0.076 = 27 day.

TV is selling in a certain area and average failure rate is 1%/2000 h. If 100 TV units are sampling and testing for 2000 h, find the probability that no accidents, $x = 0$, will occur.

$$m = n \cdot \lambda \cdot t = 100 \times 0.01/2000 \times 2000 = 1 \tag{3.30}$$

Because no accident, the probability is

$$X = 0, \, f(0) = \frac{(1)^0 e^{-1}}{0!} = 0.36 \tag{3.31}$$

3.4 Statistical Distributions

We can estimate the confidence level is 63% for 100 TV units. If no accidents, $x = 0$, keep and the confidence level would like to increase to 90%, how many TV units will it requires?

$$X = 0, \quad f(0) = \frac{(m)^0 e^{-m}}{0!} = 0.1 \tag{3.32}$$

So if $m = 2.3$, the required sample size $n = 230$ will be obtained as

$$m - n \cdot \lambda \cdot t = n \times 0.01/2000 \times 2000 = 2.3 \tag{3.33}$$

3.4.2 Exponential Distributions

The exponential distribution has a widely used application in reliability engineering because many engineering modules exhibit constant failure rate during the product lifetime. Also, it is relatively easy to handle in performing reliability analysis.

From Eq. (3.25), let $X = 0$. We can also obtain the reliability function

$$R(t) = P(0, m) = \frac{(m)^0 e^{-m}}{0!} = e^{-m} = e^{-\lambda t} \tag{3.34}$$

So the CDF also is obtained as:

$$F(t) = 1 - e^{-\lambda t} \tag{3.35}$$

If the CDF is differentiated, the PDF is obtained as:

$$f(t) = \lambda e^{-\lambda t} \quad t \geq 0, \ \lambda > 0 \tag{3.36}$$

Hazard rate function $\lambda(t)$ is defined by:

$$\lambda(t) = f(t)/R(t) = \lambda e^{-\lambda t}/e^{-\lambda t} = \lambda \tag{3.37}$$

In general, if product follows the exponential distribution, mean time to failure (MTTF) is 0.63 at $1/\lambda$ (Fig. 3.12).

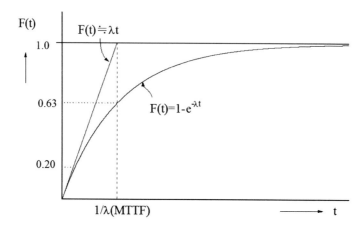

Fig. 3.12 Cumulative distribution function $F(t)$ of exponential distribution

3.5 Weibull Distributions and Its Applications

3.5.1 Introduction

In characterizing the failure times of certain components one often employs the Weibull distribution. As it was developed by Weibull in the early 1950s, this distribution can be used to represent many different failure behaviors. Many other extensions of the Weibull distribution have been proposed to enhance its capability to fit diverse lifetime data since 1970s.

This is mainly due to its weakest link properties, but other reasons have its increasing failure rate with component age and the variety of distribution shapes. The increasing failure rate accounts to some extent for fatigue failures. The density function depend upon the shape parameter β. For low β values ($\beta < 1$), the failure behavior can be similar to the exponential distribution. For $\beta > 1$, the density function always begins at $f(t) = 0$, reaches a maximum with increasing lifetime and decreasing slowly again.

As seen in Fig. 3.13, the PDF for two-parameter distribution is defined as:

$$f(t) = \frac{\beta t^{\beta-1}}{\eta^\beta} e^{-\left(\frac{t}{\eta}\right)^\beta}, \quad t \geq 0, \, \eta > 0, \, \beta > 0, \tag{3.38}$$

where η and β are characteristic life and shape parameters, respectively.

When Eq. (3.38) is integrated, the CDF is obtained as:

$$F(t) = \int_0^\infty f(t) dt = 1 - e^{-\left(\frac{t}{\eta}\right)^\beta}, \quad t > 0 \tag{3.39}$$

3.5 Weibull Distributions and Its Applications

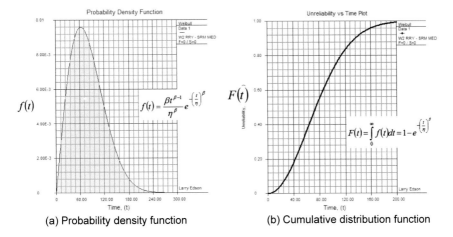

Fig. 3.13 Probability density and cumulative distribution function on the Weibull distributions

Reliability function $R(t)$ is defined as:

$$R(t) = 1 - F(t) = e^{-\left(\frac{t}{\eta}\right)^{\beta}}, \quad t > 0 \tag{3.40}$$

Hazard (or failure) rate function $\lambda(t)$ can be described by:

$$\lambda(t) = \frac{f(t)}{R(t)} = \frac{\beta}{\eta}\left(\frac{t}{\eta}\right)^{\beta}, \quad t > 0 \tag{3.41}$$

For $\beta = 1$ and 2, the exponential and Rayleigh distributions are especially called in Weibull distribution, respectively. The various failure rates of the Weibull distribution specified in bathtub curve can be divided into three regions.

- $\beta < 1.0$: Failure rates decrease with increasing lifetime (early failure)
- $\beta = 1.0$: Failure rates are constant
- $\beta > 1.0$: Failure rates increase with increasing lifetime.

For the time $\beta = 1.0$ and $t = \eta$, the CDF $F(t)$ from Eq.(3.39) can be calculated:

$$F(t) = 1 - e^{-1} = 0.632 \tag{3.42}$$

Therefore, the characteristic lifetime η is assigned to the CDF $F(t) = 63.2\%$ for exponential distribution.

3.5.2 Shape Parameters β

A shape parameter estimated from the data affects the shape of a Weibull distribution, but does not affect the location or scale of its distribution. The spread of the shape parameters represents the confidence intervals and a dependency of the stress level. A summary of the determined shape parameters is approximately described as:

- High temperature, high pressure, high stress: $2.5 < \beta < 10$
  - Low cycle fatigue: depend on cycle times
    e.g., disk, shaft, turbine
- Low temperature, low pressure, low stress: $0.7 < \beta < 2$
  - Degradation: depend on use time
    e.g., electrical appliance, pump, fuel control value.

Shape parameter β of a certain component would be invariable, but its characteristic life η varies according to use condition and material status. Thus, shape parameter (β) would be estimable and then will be confirmed after test. The density function and hazard rate function for the Weibull distribution range from shape parameters $\beta \approx 1.0$–5.0.

3.5.3 Confidence Interval

In statistics, a confidence interval (CI) is characterized as the probability that a random value lies within a certain range. CI is represented by a percentage. For example, a 90% confidence interval implies that in 90 out of 100 cases, the observed value falls within this certain interval. After any particular sample is taken, the population parameter is either in the interval realized or not. The desired level of confidence is set by the researcher. A 90% confidence interval reflects a significance level of 0.1.

The average of failure times can often deviate within a certain range. The Weibull line may describe experimental results. If the median is used to determine $F(t_i)$, 50% of the experimental results lie below the Weibull line. To know the truth of the Weibull line, it is necessary to determine its confidence interval.

Over an observation of several test samples, the Weibull line drawn in Fig. 3.14 is the most probable in the middle—median values and its confidence intervals. The line in the middle represents the population mean—observed over several test specimens—thus 50% of the cases lie above and 50% lie below this line.

3.5 Weibull Distributions and Its Applications

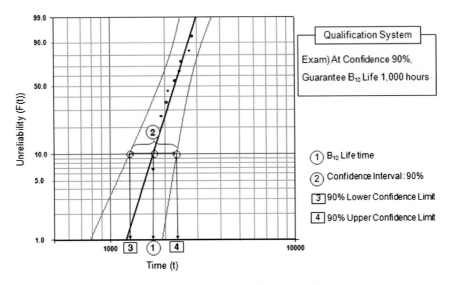

Fig. 3.14 Weibull plot of five failures with 90% confidence interval

3.5.4 A Plotting Method on Weibull Probability Paper

Weibull plotting is a graphical method for informally checking on the assumption of Weibull distribution model and also for estimating the two Weibull parameters—shape parameter and characteristic life. The method of Weibull plotting is illustrated both for complete samples of failure times (type I) or for censored samples (type II).

The CDF $F(t)$ has an S-like shaped curve (Fig. 3.15a). With a Weibull Probability Paper, If plotted the function $F(t)$ in Weibull Probability Paper, it is useful to evaluate the lifetime of mechanical component in reliability testing.

After taking inverse number and logarithmic transformation from Eq. (3.40), it can be expressed as:

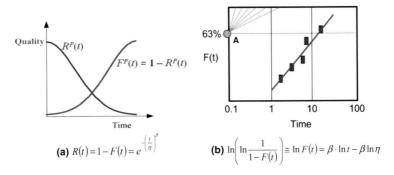

Fig. 3.15 A plotting of Weibull probability paper

$$\ln(1 - F(t))^{-1} = \left(\frac{t}{\eta}\right)^{\beta} \qquad (3.43)$$

After taking logarithmic transformation one more time, it can be expressed as:

$$\ln\left(\ln\frac{1}{1-F(t)}\right) = \beta \cdot \ln t - \beta \ln \eta \qquad (3.44)$$

If F is sufficiently small, then Eq.(3.44) can be modified as:

$$\ln\left(\ln\frac{1}{1-F(t)}\right) \cong \ln F(t) = \beta \cdot \ln t - \beta \ln \eta \qquad (3.45)$$

That is, two parametric Weibull distribution can be expressed as a straight line on the Weibull Probability Paper. The slope of its straight line becomes the shape parameter β (see Fig. 3.15b).

3.5.5 Probability Plotting for the Weibull Distribution

One method of calculating the parameters of the Weibull distribution is by using probability plotting. These procedures that are more illustrated are as following:

Step (1) Rank the times to failure in according to ascending order $t_1 < t_2 \ldots < t_n$

i	1	2	3	...	$r-1$	r
t_i	t_1	t_2	t_3		t_{r-1}	t_r

By ordering the failure times, an overview is won over the timely progression of the failure times. In addition, the ordered failure times are required in the next analysis step and are referred to as order statistics. Their index corresponds to their rank.

Step (2) Determine the failure probability $F(t_i)$ of the individual order statistics

$$F(t_i) \approx \frac{i - 0.3}{n + 0.4} \times 100 \qquad (3.46)$$

Step (3) Enter the coordinate $(t_i, F(t_i))$ in the Weibull probability paper

Step (4) Approximate sketch the best fit straight line through the entered points and determine the Weibull parameters $\hat{\beta}$. At the $F(t) = 63.2\%$ ordinate point, draw a straight horizontal line until this line intersects the fitted

3.5 Weibull Distributions and Its Applications

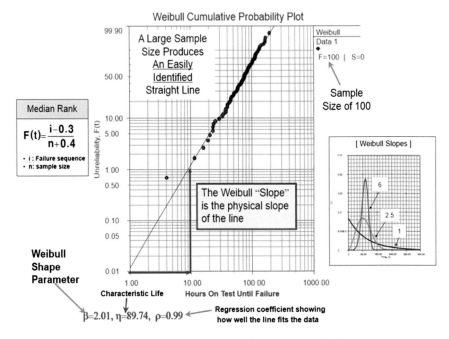

Fig. 3.16 A typical characteristics of Weibull plot with a large sample size

straight line. Draw a vertical line through this intersection until it crosses the abscissa. The value at the intersection of the abscissa is the estimate of $\hat{\eta}$ (see Fig. 3.16).

Example 3.2. Assume that six automobile units are tested. All of these units fail during the test after operating the following number of hours t_i: 93, 34, 16, 120, 53, and 75. Estimate the values of the parameters for a two-parameter Weibull distribution and determine the reliability of the units at a time of 15 h.

Solution

First, rank the times to failure in ascending order as shown next.

Failure order (i)	Time to failure (t_i) (h)	$F(t_i)$ (%)
1	16	10.94
2	34	26.56
3	53	42.19
4	75	57.81
5	93	73.44
6	120	89.06

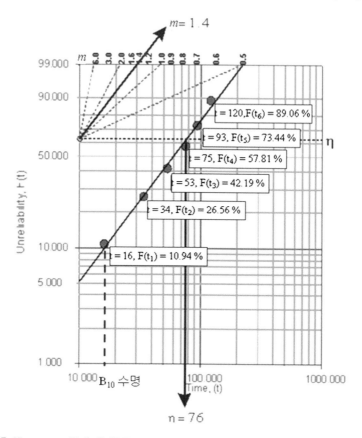

Fig. 3.17 How to use Weibull CDF

Second, by using Excel, approximate sketch the best fit straight line through the entered points $(\ln(t_i), \ln[-\ln(1 - F(t))])$ from Eq. (3.45).

Failure order (i)	$\ln(t_i)$	$\ln[-\ln(1 - F(t))]$
1	2.77	−2.16
2	3.53	−1.18
3	3.97	−0.60
4	4.32	−0.15
5	4.53	0.28
6	4.79	0.79

3.5 Weibull Distributions and Its Applications

We can obtain the estimated shape parameter $\hat{\beta}$ = slope = 1.427, estimated characteristic life $\hat{\eta}$ ($Q(t)$ is 63.2% ordinate point) = $e^{\frac{6.187}{1.427}}$ = 76.3226 h where $\ln[-\ln(1 - 0.63)] = -0.00576$, $-0.00576 = 1.427x - 6.186$.

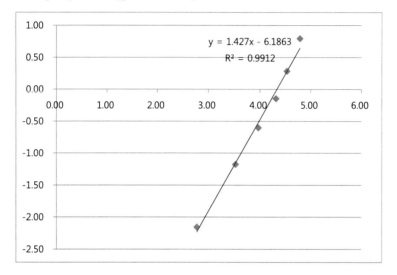

A Weibull distribution with the shape parameter β = 1.427 and η = 76.32 h is drawn on the Weibull Probability Paper. The CDF is described as

$$F(t) = 1 - e^{-\left(\frac{t}{76.32}\right)^{1.43}} \qquad (3.47)$$

In result a straight line is sketched with slope β = 1.4 on the Weibull Probability Paper. The characteristic lifetime is 76.0 h when the CDF, $iF(t)$ is 63% (see Fig. 3.17).

Reference

1. Stadterman et al. TJ (1995) Statistical field failure based models to physics of failure-based models for reliability assessment of electronic packages. Adv Electron Packaging ASME EEP 10–2:619–625

Chapter 4
Failure Mechanics, Design, and Reliability Testing

Abstract This chapter will discuss the (qualitative) established methods like FMEA, FEA, and Taguchi method that will improve the product design or quality. On the other hand, quantitative methods like Weibull analysis, reliability testing and the others also will be discussed further as a methodology of reliability assessment. As time goes, product becomes failure-the state of not meeting an intended function of the customer's satisfaction. Product failures in field happen when the parts cannot withstand the repetitive stresses due to loads over the product lifetime. The failure mechanics of product can be characterized by the stress (or loads) on the structure and materials used in the structure. If there is a void (design weak point) in the material where the loads are applied, engineer would want to move the void in the structure to location away from where the stress is applied. These activities are called design. Quantitatively, the final goal of these quality activities is to discover the design problems by reliability testing. Engineer judges whether the product will achieve the reliability target.

Keywords Reliability testing · Failure mechanics · Design · Quantitative method

4.1 Introduction

Reliability can be defined as the probability that a component or product will fulfill its intended function over lifetime. It is necessary to clearly understand what a product's intended function and its failure are. Intended functions are the product functionalities that perform the voices of the customer. They are represented as product specifications in company. As time goes, product becomes the state of not meeting an intended function of the customer's satisfaction. Consequently, we call it as failure (Fig. 4.1).

The most common failures are those caused by specification deficiencies of the product function. In such cases design might result in the product failure to customer satisfaction (or specification). When field failure occurs, we also determine whether the company specifications are inappropriate or whether verifiers are

Fig. 4.1 Train wreck or train crash from Wikipedia

incorrectly conforming to the specifications. To find out the design faults, verification specifications suitable for a newly developed product should be developed like reliability quantitative specifications that are described in Chap. 7.

If functions break during product usage unexpectedly, we can say it is failure. The definition of a failure may not be precise if a gradual or intermittent loss of performance over time is observed. For an example, seals experience a degradation of material properties and no longer satisfy the specifications instantaneously. In this case we can replace the old part with a new one. The most critical failures are no longer satisfied with the customer requirements (or specifications) due to unidentified factors after the product releases. In such cases design might does not achieve the customer requirements and experience the recall. Thus, design is a critical process to determine whether the intended function of product or modules in lifetime fulfills through exact specifications.

A disaster due to product failure is always an undesirable event for several reasons: putting human lives in jeopardy, causing economic losses and interfering with the availability of products and services. The failure causes come from improper materials selection, inadequate design of the parts and its misuse. So it is the engineer's responsibility to be prepared when failure is expected to occur. Engineer needs to assess its cause and take action and appropriate preventive measures against future incidents—the parametric ALT will be discussed in Chap. 7.

4.2 Failure Mechanics and Designs

The failure mechanics of components that might no longer be functioned can be characterized by two factors: (1) the stress (or loads) on the structure, (2) The type of materials used in the structure. To prevent the failure, engineer should know either loads (or stress) or structural (or materials) related (Fig. 4.2).

If there is a void (design weak point) in the material where the loads are applied, the structure can fracture at that location. The engineer would want to move the void in the structure to a location away from where the stress is applied.

Product failure in mechanical system is a physical problem that is created when stress due to loads causes a fracture. Failure mechanics seeks to understand the process how stress and materials impact the failure. The applied loads cause stresses on the module structure. The failure site of the module structure might be observable when the failed products are taken apart in the field or could be from the results of parametric ALT.

If the structure is ideally designed and has well-dispersed stresses, there should be no problems with the failure of the module. As the mechanical design is developed with an optimal design process-Finite Element Analysis (FEA), it may have design flaws that will show up in the field. Figure 4.3 illustrates an ideal design process that includes design, reliability testing, and field conditions feeding back into the design and reliability testing. Product failures in the field happen when the parts cannot withstand the repetitive stresses due to loads over the life time of the module. Most products with electrical or mechanical components are composed of multi-module structures. If one of the modules has a problem due to an improper design, then that module will determine the lifetime of the failed product. Consequently, by experiments such as reliability testing, the design problems might

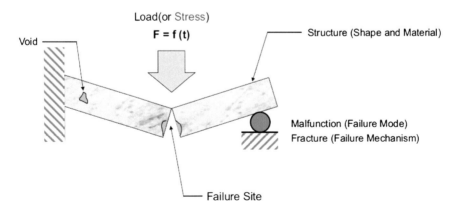

Fig. 4.2 Failure mechanics created by a load on a component made from a specific material

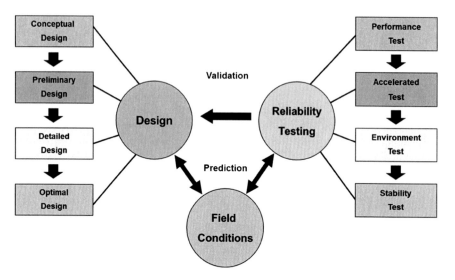

Fig. 4.3 Design process and reliability testing

be revealed before launching product. The product design should be effectively connected with Design Engineering and Test Engineering to achieve the reliability target of modules.

4.2.1 Product Design—Intended Functions

We can say to define the intended functions that shall implement the customer's needs in product. For television, the intended function is to watch the program that consists of moving images. It is required as the fundamental advantage like the superiority of picture and sound. A product like TV consists of multiple modules that can be put together as a subassembly. When product has an input, it has output as response (intended functions) that want to be implemented. Intended functions are embodied in the product design process and their performance will be measured by specifications (See Fig. 4.4).

Based on design idea, intended functions of product can be implemented through the design developing process. In the same manner designers use the engineering design process to satisfy the customer's needs. Product design would learn from the experiences like previous market failure. Companies often specify the past mistakes by the documents that describe the design requirements. Performance are to note each of the key features that consist of a variety of specification. Reliability is to identify the failure mode of product through testing over time or lifetime. Manufacture determines whether product conforms to specification in the production process. After production, the new concept of product determines the feedback of customer in field (Fig. 4.5).

4.2 Failure Mechanics and Designs

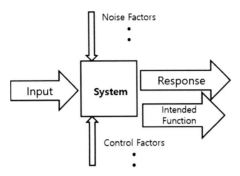

Fig. 4.4 Intended function in parameter diagram

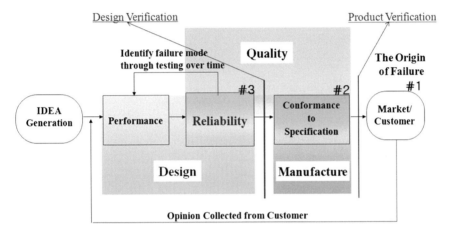

Fig. 4.5 Implementing intended functions in the design process

There are many good possibilities for solving design problems and choosing the best solution that product specifications could be embodied in the design process. Product development refines and improves a solution in the design phase before a product launches to customers. Basically, the engineering design process starts to define the problems: (1) What is the problem to be implemented? (2) How have others approached it? (3) What are the design constraints?

A prototype is an first operating version of a solution (or intended function) that satisfies the customer requirements. The product design process involves multiple iterations like redesigns of your solution before settling on a final design. Final product can define the performance criteria in evaluating the specifications of product quality—(a) intended functions (or fundamental advantage), (b) specified product life, and (c) the failure rate of product under operating and environmental conditions.

In the current global competitive marketplace, product quality of intended functions is an important requisite to ensure continued success in the marketplace.

On the other hand, the failure of product quality expels from the customer satisfaction and loyalty. Thus, it is important for the product design team to understand customer expectation or voice and usage conditions.

Intended functions are the functionalities that product is to perform based on the voices of the customer and explain the company specifications. The intended functions of a product must be recognized in the design to ensure whether critical customer requirements (or specifications) work properly in product. Thus, the fulfillment of each intended function should be understood from a standpoint of the customer's expectations. As time goes by, product fails to perform its intended functions because they don't meet. For example, in the first steps of failure mode effects analysis (FMEA), a product starts to assess whether a product might fail to perform its intended function.

The engineering design team may often neglect the potential customer uses, though customer's right uses. Sometimes the failure of a returned product in field may be perceived to be customer abuses of product, generally not failure. From a standpoint of robust design, the intended functions might be designed to withstand the customer misusages (or overloads) for cusotmer's proper uses. In this case the robust concept is effective. To robustly keep the intended functions, product withstands noise factors like customer usage (or loads). As seen in Fig. 4.4, Taguchi's robust design schematic of product employs two experimental arrays: one for the control array (design) and the other for the noise array (loads). Optimizing over the control factors, product can be reduced to a signal (output)-to-noise (load) ratio, and intended function will be designed robustly.

However, a large number of experimental trials in the Taguchi product array may be required because the noise array is repeated for every row in the control array. As a result it is hard to discover the optimal design parameters. Alternative approach for a robust design of intended function will explain the parametric ALT in Chap. 7.

4.2.2 Specified Design Lifetime

The design life of product is the time period which the product works properly within the life expectancy. The design life of products might differ from the reliability lifetime metrics—MTBF. For example, the MTBF of product may be 100,000 h and the design life is 10,000 h. It means that one failure occur every 100,000 population operating hours. Because product cannot reach 100,000 operating hours, most of these units will be replaced by a new unit. Aluminum electrolytic capacitors, fans, and batteries will fail due to wear-out before they could achieve the operating time—MTBF. As design life time, BX life can be useful. It means the life time at which $X\%$ of the units in a population will have failed. For example, if unit has a B10 life of 10,000 h, 10% of the population will have failed by 10,000 h of operation.

The specified design lifetime provides a usage or timeframe for reliability analysis or testing. Some organizations might simply choose to design a product to

4.2 Failure Mechanics and Designs

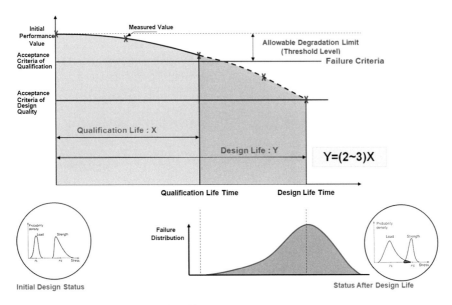

Fig. 4.6 Qualification life and design life

be reliable over the stated warranty (qualification) period. Because it is commensurate with how long the product is expected to be used in the field, an enlightened organization might choose a design life. Depending on the designer's perspective, the life specification might be based on any of the design life time—BX life.

When product is selling to customer, product is designed to have safe margin between environmental stress and strength. As time goes on, product failure initially appears at qualification lifetime. Design life has maximum failure rate at acceptance criteria of design quality. Generally, design lifetime is 2–3 times greater than qualification lifetime. Design life is based on information coming from the customer and competitive benchmarks. Engineers often define a design life time to represent an engineer's specifications of product usage under which the reliability must be verified. It is important that careful thought go into the synthesis of this specification to ensure that product quality do not result in customer's dissatisfaction due to the end of the useful life (Fig. 4.6).

4.2.3 Dimensional Differences Between Quality Defects and Failures

As seen in Table 4.1, quality defects appear for one of three reasons: (1) Incompleteness of design specification, (2) Nonconformance to specifications during manufacture, (3) Customer misuse of the product. The established specifications in company can perceive whether product quality is met. They also will follow the normal distribution.

Table 4.1 Quality defects and failures

	Quality defects	Failure
Concept	Out of established specifications	Physical trouble
Index	Defect rate, ppm	Failure rate, lifetime
Unit	Percent, ppm	Percent/year, year
Area	Manufacturing	Design
Probability	Normal distribution $f(x) = \frac{1}{\sigma\sqrt{2\pi}} e^{-\frac{(x-\mu)^2}{2\sigma^2}}$	Exponential/Weibull $F(t) = 1 - R(t) = 1 - e^{-\lambda t}$

The quality defects can refer to a variety of problems that appear before customer use, such as flaws causing poor performance or the failure to work. Similarly, the good quality means good aesthetic design with fundamental advantage as well as good performance. For some aspect of the product, quality defect is out of tolerance in manufacturing. If something is found to be out of specifications, it can be considered to have a quality defect that follows the normal distribution.

On the other hand, failure clearly indicates a physical performance disorder related to the product as time goes on. The number of failure per year for a given production lot results in the annual failure rate. Any mistakes or omissions of design will induce results in failure that follows the exponential or Weibull distributions. The absence of failure means good reliability.

4.2.4 Classification of Failures

The definition of failure is obvious when there is a total loss of product (intended) functions that can be differentially perceived from the viewpoints of the customers or by specifications. If something breaks during product usage unintentionally, it may fail. However, if only a partial loss of (intended) function is involved, it will be complicated to define the product failure. In such instances the definition of the failure may not be precise when one observes a gradual or intermittent loss of performance over time. Although the activity is completed successfully, a person may still feel dissatisfied if the underlying process is perceived to be below expected standard (or specification).

4.2 Failure Mechanics and Designs

For example, a variety of automobile failures might be classified as

A Class is said that failure will damage the body of passenger or the loss of car control, failure of brake equipment, and fire risk. Examples are too many to be expected: (1) Accident due to loss of acceleration control, (2) Differential gear fixation, (3) air exposure of pump, (4) Overheat or disconnection of cable, (5) Damaged flywheel, (6) Malfunctioned clutch, and (7) Malfunctioned injection pump (See Fig. 4.7).

B1 Class is said that failure will stop car. Examples are too many to be expected: (1) Engine stop or no starting from computing engine, injection, pump, common rail, ignition coil, car starting, engine control, engine fixation, distribution chain, (2) Transmission stop, and (3) Stop of gear box, no reverse operation.

B2 Class is said that failure may stop car. There are a lot of examples: (1) Abnormal noise of engine or gear box, (2) Overheat Engine, and (3) Vibration of engine or gearbox.

C Class is said that car can be drivable, but it requires the high cost to recover it. Examples also are too many to be expected: (1) It make car inoperative, (2) It affects visual, hearing, and smell, (3) Critical motor surges and power loss, (4) Abnormal noise, oil spill, cooling water spill, smell, over oil leakage, (5) abnormal smell, and (6) Clutch malfunction, inoperative gear transmission. C1 Class also is inconsistent to the standard of discharge gas. Example is not to meet for standard of emission gas.

D Class is said that using car is no effect but minor operational failure. Examples are too many to be expected: (1) Driving car is inconvenient, (2) Affect visual and hearing, (3) Over fuel consumption, (4) Slowly acceleration, (5) Idle speed is

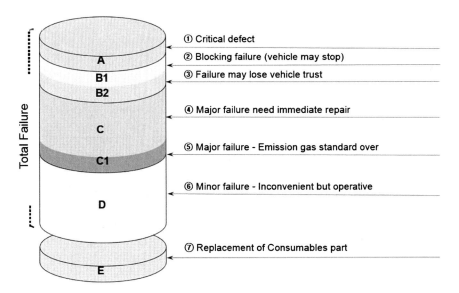

Fig. 4.7 Definition for failure class (example: automobile)

instable, (6) Engine is not starting, (7) Vibration, and (8) A few noises (discharge noise, cracking noise, vibration noise, cooling pump noise, starting noise, noise in gear transmission, erosion of engine part).

E Class is wearable parts such as filter, spark plug, and timing belt need to be replaced periodically. In a result, we can say failure is defined as A, B, C, and D class.

4.3 Failure Mode and Effect Analysis (FMEA)

4.3.1 Introduction

Failure mode and effect analysis (FMEA) is a widely used method to study system problems in the reliability engineering. The history of FMEA goes back to the early 1950s with the development of flight control systems when the U.S. Navy's Beau of Aeronautics developed a requirement called "Failure Analysis." In the mid-sixties, FMEA was set to work in earnest by NASA for the Apollo project. In the 1970s, the U.S. Department of Defense developed military standards entitled "Procedures for Performing a Failure Mode, Effect, and Critically Analysis." For use in aerospace, defense, and nuclear power generation, FMEA/FMECA methods are widely used to conduct analysis of systems. The Ford Company integrated this method into its quality assurance concept.

- In the early of 1950s: Propeller airplane ⟶ Reliability design of Jet airplane with flight control systems
- In the middle of 1960s: Apollo program used in real earnest
- In the 1970s: US NAVY (MIL-STD-1629) was adopted
- In the late of 1970s: widely used in the industry due to the introduction of product liability law

In the first step of a system reliability study, FMEA involves reviewing the design of many components, assemblies, and subsystems to identify failure modes, and their causes and effects. To determine whether an optimum criterion of reliability assessment is achieved, FMEA is to analyze and modify many components in system. FMEA uses the risk priority number (RPN). A qualitative analysis is mainly used in the light of FMEA team experience.

It may be described as an approach used to perform analysis of each potential failure mode in the systems under consideration to examine the effects of such failure modes on that system. When FMEA is extended to classify each potential failure effect according to its severity, the method is known as failure mode effects and criticality (FMECA).

As seen in Fig. 4.8, the FMEA is carried out in interdisciplinary groups—members in the planning, R&D, and QA. It is reasonable to execute an FMEA in teams, since it is only then possible to incorporate all operational areas affected by

4.3 Failure Mode and Effect Analysis (FMEA)

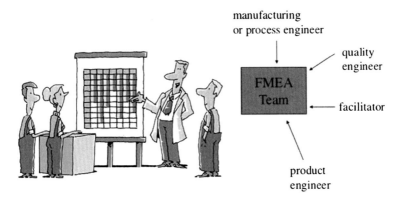

Fig. 4.8 Members of FMEA team

the analysis. In practice it is beneficial to execute an FMEA under the direction of an FMEA moderator, who is familiar with the method. In this way, time-consuming discussions concerning the method can be avoided.

In general, the FMEA team consists of a moderator, who offers methodical knowledge. They can offer technical knowledge concerning the product or process to be analyzed. The moderator, who also may possess a marginal know-how concerning the product or process, certifies that the team members acquire a basic knowledge of the FMEA methodology. A brief training at the beginning of an FMEA assignment is useful.

FMEA is a systematical method that the fundamental idea is the determination of all possible failure modes for arbitrary systems or modules and the possible failure effects and failure causes are presented. The aim of the method is to recognize the risks and weak points of product design as early as possible in order to enable execution improvements in a timely manner. There are many terms used in performing FMEA/FMECA and some of them are as follows:

- **Failure cause**. The factors such as design defects, quality defects, physical or chemical processes, or part misapplications are the primary reason for failure or they start the physical process which deterioration progresses to failure.
- **Failure mode**. The notion or manner through which a failure is perceived.
- **Failure effect**. The consequence a failure mode has on item's function, operation, or status.
- **Single failure point**. An item's malfunction that would lead to system failure and is not compensated through redundancy or through other operational mechanisms.
- **Criticality**. A relative measure of a failure mode's consequences and its occurrence frequency.
- **Severity**. A failure mode's consequences, taking into consideration the worst case scenario of a failure, determined by factors such as damage to property, the degree of injury, or ultimate system damage.

- **Criticality analysis.** An approach through which each possible failure mode is ranked with respect to the combined influence of occurrence probability and severity.
- **Undetectable failure.** A postulated failure mode in the FMEA for which no failure detection approach is available through which the concerned operator can be alerted of the failure.
- **Local effect.** The consequences a failure mode has on the function, operation, or status of the item currently being analyzed.

4.3.2 Types of FMEA

The types of FMEA are classified as (1) System-level FMEA, (2) Design-level FMEA, and (3) Process-level FMEA (See Fig. 4.9).

4.3.3 System-Level FMEA

Failure functions as well as failure modes for product are analyzed in the system-level FMEA. The analysis is carried out in various hierarchical system

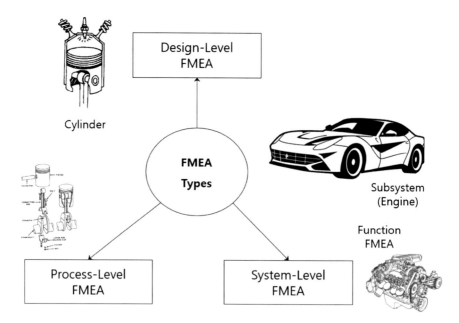

Fig. 4.9 Types of failure mode and effect analysis (FMEA)

levels all the way to the failure on the module level. This is the highest level FMEA that can be performed and its purpose is to identify and prevent failures related to system/subsystems during the early conceptual design. System-level FMEA is carried out to validate that the system design specifications reduce the risk of functional failure to the lowest level during the operational period.

Some benefits of the system-level FMEA are identification of potential systemic failure modes due to system interaction with other systems and/or by subsystem interactions, selection of the optimum system design alternative, identification of potential system design parameters that may incorporate deficiencies prior to releasing hardware/software to production, a systematic approach to identify all potential effects of subsystem/assembly part failure mode for incorporation into design-level FMEA, and a useful data bank of historical records of the thought processes as well as of action taken during product development efforts.

4.3.4 Design-Level FMEA

The purpose of performing design-level FMEA is to help identify and stop product failures related to design. This type of FMEA can be carried out upon component-level/subsystem-level design proposal and its intention is to validate the design chosen for a specified functional performance requirement. The advantages of performing design-level FMEA include identification of potential design-related failure modes at system-/subsystem-/component-level, identification of important characteristics of a given design, documentation of the rationale for design changes to guide the development of future product design, help in the design requirement objective evaluation of design alternatives, systematic approach to reduce criticality and risk, accumulated data serve as a useful historical record of the thought processes and the actions taken during the product development effort, and useful tool to establish priority for design improvement actions.

4.3.5 Process-Level FMEA

This identifies and prevents failures related to the manufacturing/assembly process for a certain product. The benefits of the process-level FMEA include identification of important characteristics associated with the process, identification of potential process shortcomings early in the process planning cycle, development of priorities for process improvement actions, and documentation of rationale for process changes to help guide the establishment of potential manufacturing processes.

4.3.6 Steps for Performing FMEA

The distinction between technical knowledge in various fields and the methodology of an FMEA execution offers the advantage that the experts from the respective fields only offer their technical knowledge free of any methodical considerations. Thus, merely a basic knowledge of FMEA is adequate of the team of experts. The team size ranges ideally between 4 and 6 members—supervisors, moderator, and a small product team.

The fundamental step of an FMEA searches for all conceivable failure modes. Thus, this step should be executed most carefully. Each failure mode that is not found can lead to dangerous failure effects. Options available to discover failure modes are damage statistics, experience of the FMEA participants, checklists, brainstorming, and systematic analysis over failure functions. An imperative principle is the observation of former arisen failures in similar cases. All further failure modes can be derived with the help of the experience of the FMEA participants.

The first section of the FMEA form sheet are reserved for the description of the system, product. or process and their function. The next section of the form sheet deals with the risk analysis. This is followed by a risk assessment in order to rank the numerous failure causes. The last step is a concept optimization derived from the analysis of the risk assessment (See Fig. 4.10).

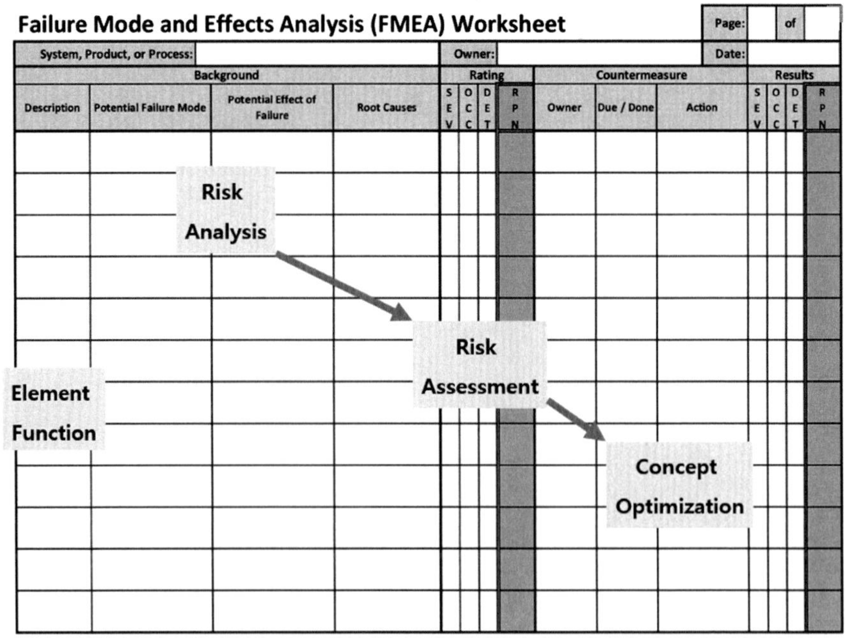

Fig. 4.10 FMEA form sheet that consists of four sections

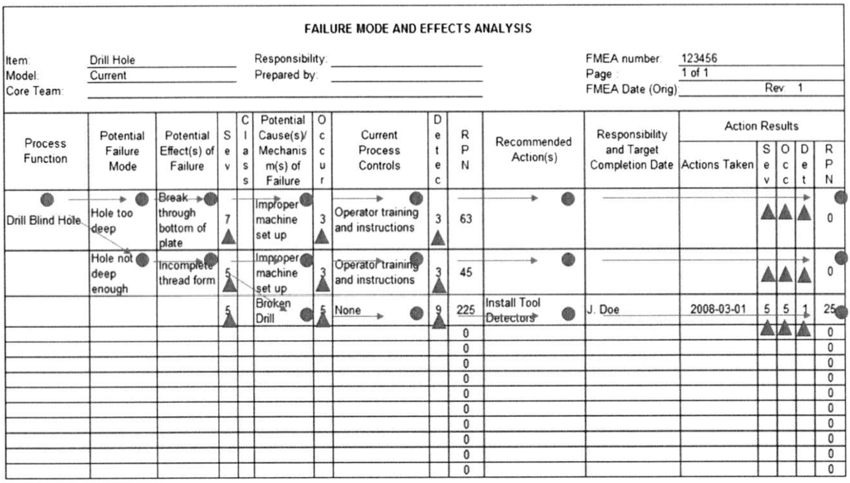

4.3 Failure Mode and Effect Analysis (FMEA)

FAILURE MODE AND EFFECTS ANALYSIS

Item: Drill Hole	Responsibility:	FMEA number: 123456
Model: Current	Prepared by:	Page: 1 of 1
Core Team:		FMEA Date (Orig): ___ Rev. 1

Process Function	Potential Failure Mode	Potential Effect(s) of Failure	Sev	Cls	Potential Cause(s)/ Mechanism(s) of Failure	Occur	Current Process Controls	Detec	RPN	Recommended Action(s)	Responsibility and Target Completion Date	Action Results				
												Actions Taken	Sev	Occ	Det	RPN
Drill Blind Hole	Hole too deep	Break through bottom of plate	7		Improper machine set up	3	Operator training and instructions	3	63							0
	Hole not deep enough	Incomplete thread form	5		Improper machine set up	3	Operator training and instructions	3	45							0
					Broken Drill	5	None	9	225	Install Tool Detectors	J. Doe	2008-03-01	5	5	1	25
									0							0
									0							0
									0							0
									0							0
									0							0
									0							0
									0							0
									0							0

Fig. 4.11 FMEA example for mechanical/civil parts with tree structure

The completed form sheet represents a tree structure. A certain component has one or more functions and normally several failure modes. Each failure mode has again various failure effects and different failure causes (See Fig. 4.11).

FMEA can be performed in the following six steps:

4.3.6.1 Defines System and Its Associated Requirements (Step1)

This is a first step of FMEA. Define the system under consideration how complex the system is. The analyst must develop the system definition using documents such as reports, drawings, development plans (or specifications). The system structure arbitrarily orders the individual system elements into various hierarchical levels.

- The definition of design (or process) interfaces
- Dividing the system into its individual system elements—module and components
- Arranging system elements hierarchically in product structure

4.3.6.2 Describe the System and Its Associated Functional Blocks (Step 2)

- The arrangement of the system structure is the basis for determining the preparation of the description of the system under consideration. Such description may be grouped into two parts.

- **Narrative functional statement** (**Top down**). The functions are created by preparing for each module and component as well as for the total system (See Fig. 4.3). It provides narrative description of each item's operation for each mode/mission phase. The degree of the description detail depends on factors such as an item's application and the uniqueness of the functions performed.
- **System block diagram**. The purpose of this block diagram is to determine the success/failure relationships among all the system components.

4.3.6.3 Identify Failure Modes and Their Associated Effects (Failure Analysis, Step 3)

A failure analysis performs the analysis and determination of the failure modes and their effects. The failure leads to the dissatisfaction of a module. Compensating provisions and Criticality classification are described below.

- **Compensating provisions**. Design provisions or operator actions that is circumventing or mitigating the failure effect.
- **Criticality classification**. This is concerned with the categorization of potential effect of failure.

  - People may lose their lives due to failure
  - Failure may cause mission loss
  - Failure may cause delay in activation
  - Failure has no effect

4.3.6.4 Risk Assessment (Step 4)

The objective of the risk assessment is to prioritize the failure modes discovered during the system analysis on the basis of their effects and occurrence likelihood. Thus, for making an assessment of the severity of an item failure, two commonly used methods are Risk Priority Number (RPN) Technique that is widely used in the automotive industrial (Tables 4.2, 4.3 and 4.4).

4.3.6.5 RPN (Risk Priority Number)

This method calculates the risk priority number for apart failure mode using three factors: (1) failure severity ranking (SR), (2) failure mode occurrence ranking (OR), and (3) failure detection probability (DR). For example, if people are put into danger, the severity is evaluated higher, whereas a minimum limitation of comfort would receive a respectively lower value. With assessment value DR it is determined how successful the detection of the failure cause is before delivery to the

4.3 Failure Mode and Effect Analysis (FMEA)

Table 4.2 Failure detection ranking

Item no.	Likelihood of detection	Rank meaning	Rank
1	Very high	Potential design weakness almost certainly detected	1, 2
2	High	There is a good chance of detecting	3, 4
3	Moderate	There is a possibility of detecting potential design weakness	5, 6
4	Low	Potential design weakness is unlikely to be detected	7, 8
5	Very low	Potential design weakness probably will not be detected	9
6	Delectability Absolutely uncertain	Potential design weakness cannot be detected	10

customer. More specifically, the risk priority number is computed by multiplying the ranking (i.e., 1–10) assigned to each of these three factors. Thus, the risk priority is expressed by

$$\text{Risk Priority Number (RPN)} = (\text{Severity})(\text{Occurrence})(\text{Severity}) \quad (4.1)$$

With the RPN a ranking of the identified failure causes and their failure connection to the failure effect can be done.

Since the above three factors are assigned rankings from 1 to 10, the value of the RPN will vary from 1 to 1000. The average RPN is normally 125 ($5 \times 5 \times 5$). Failure modes with a high RPN are considered to be more critical; thus, they are given a higher priority in comparison to the ones with lower RPN. Nonetheless, ranking and their interpretation may vary from one organization to another. Table 4.3 uses three present rankings for failure detection, failure mode occurrence probability, and failure effect severity in one.

Table 4.3 Failure mode occurrence probability

Item no.	Ranking term	Rank meaning	Occurrence probability	Rank
1	Remote	Occurrence of failure is quite unlikely	<1 in 10^6	1
2	Low	Relatively few failures are expected	1 in 20,000 1 in 4000	2 3
3	Moderate	Occasional failures are expected	1 in 1000 1 in 400 1 in 80	4 5 6
4	High	Repeated failures will occur	1 in 40 1 in 20	7 8
5	Very high	Occurrence of failure is almost inevitable	1 in 8 1 in 2	9 10

4.3.6.6 Optimization (Step 5)

The last phase of the FMEA is the optimization phase. First, the calculated RPN are ordered according to their values. According to the Pareto principle, 20–30% of the RPN has been optimized.

- Ranking of failure causes according to their RPN value
- Concept optimization beginning with the failure causes with the highest RPN (Pareto principle)
- Failure causes with OR > 8, DR > 8, DR > 8 separately

The new optimization actions are entered on the right side of the form sheet for the optimized failure causes and the responsibility is recorded. An improved RPN is calculated for the improved state the new assessment values assigned to RPN.

Example 4.1 Develop a FMEA for pressure cooker like Fig. 4.12. The safety features of pressure cooker are the following:

(1) Safety valve relieves pressure before it reaches dangerous levels.
(2) Thermostat opens circuit through heating coil when the temperature rises above 250 °C.

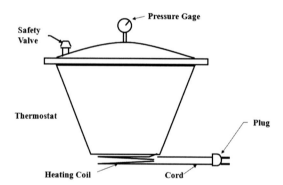

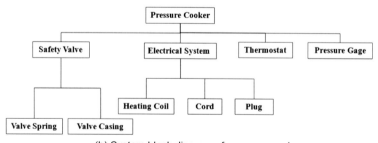

Fig. 4.12 FMEA for pressure cooker

4.3 Failure Mode and Effect Analysis (FMEA)

Table 4.4 Severity of the failure mode effect

Item no.	Failure effect severity category	Severity category description	Rank
1	Minor	No real effect on system performance and the customer may not even notice the failure	1
2	Low	The occurrence of failure will only cause a slight customer annoyance	2, 3
3	Moderate	Some customer dissatisfaction will be caused by failure	4, 5, 6
4	High	High degree of customer dissatisfaction will be caused by failure but the failure itself does not involve safety or noncompliance with government rules and regulations	7, 8
5	Very high	The failure affects safe item operation, involves noncompliance with government rules and regulations	9, 10

(3) Pressure gauge is divided into green and red sections. That is, "Danger" is indicated when the pointer is in the red section.

First of all, problem should define scope:

(1) Resolution: The analysis will be restricted to the four major subsystems (electrical system, safety valve, thermostat, and pressure gauge)
(2) Focus—Safety

Based on the focus of safety of pressure cooker, perform Failure Modes, Effects, and (Criticality) Analysis for a Pressure Cooker with

4.4 Fault Tree Analysis (FTA)

4.4.1 Concept of FTA

Fault tree analysis (FTA) is one of the most widely used methods in the industrial area to identify the internal (or external) causes of failures (Fig. 4.13). Thus, the FTA defines the system behavior in regard to fault. FTA was developed in the early 1950s at Bell Telephone Laboratories and started to use the FTA for the development of commercial aircrafts (1966). In the 1970s this method was used in the area of nuclear power. Now it is spread in many different areas—automobile, communication, and robotics.

FTA is used to show the system functions and their reliability. In the early design stage it may be applied as a diagnosis and development tool. The potential system faults can be identified and the design action plans can be setup. One of the major advantages of FTA is that the method provides both qualitative and

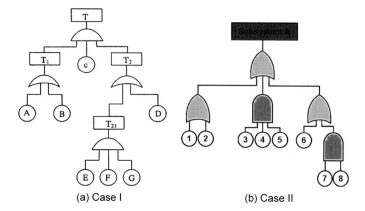

Fig. 4.13 A typical example of fault tree analysis

Fig. 4.14 Commonly used fault tree symbols: (i) AND gate, (ii) OR gate, (iii) rectangular, (iv) OR gate

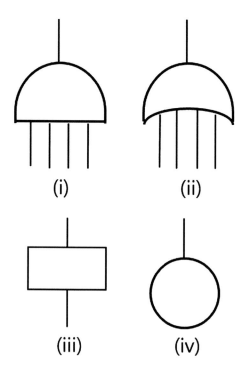

quantitative results.FTA also with Boolean algebra and probability theory is beneficial to the preventative quality assurance (See Fig. 4.14).

Although many symbols are used in performing FTA, the four commonly used symbols are described as

4.4 Fault Tree Analysis (FTA)

- **AND gate**. This denotes that an output fault event occurs only if all of the input fault events occur
- **Or gate**. This denotes that an output fault event occurs only if one or more of the input fault events occur
- **Rectangle**. This denotes a fault event that results from the logical combination of fault events through the input of a logic gate
- **Circle**. This represents a basic fault event or the failure of an elementary component. The event's probability of occurrence, failure, and repair rates are normally obtained from field failure data.

The objectives of FTA are (1) systematic identification of all possible failures and their causes, (2) illustration of critical failures, (3) evaluation of system concepts, and (4) documentation of the failure mechanism and their functional relations. It begins by identifying an undesirable system event (Top Event). Top event are generated and connected by logic gates such as OR and AND. The fault tree constructions are repeated successively until the lowest events are developed.

Example 4.2 Assume that electric circuit system contains motor system, two switches, and electric power source. Develop a fault tree for the top event "no operating motor," if the interruption of motor power can only be caused either by current failure or motor failure (See Fig. 4.15).

By using the Fig. 4.14 symbols', a fault tree for motor system can be developed as follows:

Each fault event in the figure is labeled as X_0, X_1, X_2, X_3, X_4, X_5, X_6, X_7, and X_8. For independent fault events, the probability of occurrence of top events of fault trees can easily be evaluated by applying the basic rules of probability to the output fault events of logic gates (Fig. 4.15). For example, we have

$$P(x_4) = P(x_7) + P(x_8) - P(x_7)P(x_8) \tag{4.2}$$

$$P(x_1) = P(x_3) + P(x_4) - P(x_3)P(x_4) \tag{4.3}$$

$$P(x_2) = P(x_5) + P(x_6) - P(x_5)P(x_6) \tag{4.4}$$

$$P(x_0) = 1 - [1 - P(x_1)][1 - P(x_2)] \tag{4.5}$$

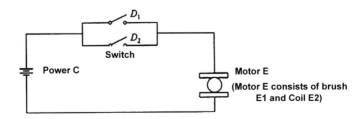

Fig. 4.15 A circuit diagram for Example 4.2

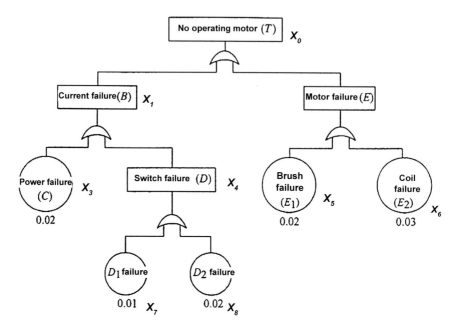

Fig. 4.16 A fault tree for Example 4.1

where $P(X_i)$ is the probability of occurrence of fault event X_i, for $i = 1, 2, 3, \ldots, 8$.

Example 4.3 In Fig. 4.16, assume that the probability of occurrence of fault events X_3, X_5, X_6, X_7, and X_8 are 0.02, 0.02, 0.03, 0.01, and 0.02, respectively. Calculate the probability of occurrence of the top event "no operating motor" using Eqs. (4.2)–(4.5).

Thus, by substituting the given data values into Eqs. (4.2)–(4.5), we can get

$$P(x_4) = P(x_7) + P(x_8) - P(x_7)P(x_8) = (0.01) + (0.02) - (0.01)(0.02) = 0.0298 \tag{4.6}$$

$$P(x_1) = P(x_3) + P(x_4) - P(x_3)P(x_4) = (0.02) + (0.0298) - (0.02)(0.0298)$$
$$= 0.0492 \tag{4.7}$$

$$P(x_2) = P(x_5) + P(x_6) - P(x_5)P(x_6) = (0.02) + (0.03) - (0.02)(0.03) = 0.0494 \tag{4.8}$$

$$P(x_0) = 1 - [1 - P(x_1)][1 - P(x_2)] = 1 - (1 - 0.0492)(1 - 0.0494) = 0.0962 \tag{4.9}$$

Thus, the probability of occurrence of the top event "no operating motor" is 0.0962

4.4.2 Reliability Evaluation of Standard Configuration

Engineering systems can form various types of configurations in performing reliability analysis. A system is said to be a serial system if failure of one or more components within system results in failure of the entire system. On the other hands, parallel system is that the failure of all components within the system results in the failure of the entire system. For example, the lighting system that consists of four bubs in a room is a parallel system, because room blackout happens only when all four bulbs break. The reliabilities the serial or parallel systems are summarized in Table 4.5.

Example 4.4 Assume that an aircraft has four independent and identical engines and all must work normally for the aircraft to fly successfully. Calculate the reliability of the aircraft flying successfully, if each engine's reliability is 0.99.

Table 4.5 Failure modes, effects, and (criticality) analysis for a pressure cooker

Item	Failure mode	Failure causes	Failure effects	S	O	C	Control measures/remarks
Electrical system	No current	Defective cord Defective plug Defective heating coil	Cooking interruption (mission failure)	1	2	2	Use high-quality components Periodically inspect cord and plug
	Current flows to ground by an alternate route	Faulty Insulation	Shock Cooking interruption	2	1	2	Use a grounded (3-prong) plug. Only plug into outlets controlled by ground-fault circuit interrupters
Safety valve	Open	Broken valve spring	Steam could burn operator Increased cooking time	2	2	4	Design spring to handle the fatigue and corrosion that it will be subjected to
	Closed	Corrosion Faulty manufacture	Potential over pressurization	1	2	2	Use corrosion-resistant materials Test the safety valve
Thermostat	Open	Defective thermostat	Cooking interruption	1	2	2	Use a high-quality thermostat
	Closed	Defective thermostat	Over pressurization eventually opens valve	1	2	2	Use a high-quality thermostat

By substituting the given data values of equation system reliability in Table 4.6, we can get

$$R_s = (0.99)^4 = 0.9606 \tag{4.10}$$

Thus, the reliability of the aircraft flying successfully is 0.9606 (Fig. 4.6).

Example 4.5 A system is composed of two independent and identical active units and at least one unit must operate normally for the system success. Each unit's constant failure rate is 0.0008 failures per hour. Calculate the system mean time to failure and reliability for a 150-hour mission.

Substituting the given data values of parallel system MTTF Eq. in Table 4.6 yields

Table 4.6 Standard Configuration with m units

System structure	Serial systems	Parallel systems
Block diagram	X_1 — X_2 ⋯ X_m	X_1, X_2, X_m in parallel
Functional tree	AND gate with $X_1 \ldots X_m$ → Y	OR gate with $X_1 \ldots X_m$ → Y
System reliability	$R_s(t) = \prod_{i=1}^{m} R_i(t)$	$R_s(t) = 1 - \prod_{i=1}^{m}(1 - R_i(t))$
ith unit constant failure rate	$R_i(t) = e^{-\int \lambda_i dt} = e^{-\lambda_i t}$	$R_{ps} = 1 - \prod_{i=1}^{m}(1 - e^{\lambda i})$
MTTF	$\int_0^\infty R_i(t)dt = 1/\sum_{i=1}^{n} \lambda_i$	$\int_0^\infty R_i(t)dt = \frac{1}{\lambda}\sum_{i=1}^{n}\frac{i}{i}$

4.4 Fault Tree Analysis (FTA)

$$\text{MTTF}_{ps} = \int_0^\infty \left[1 - \left(1 - e^{-\lambda t}\right)^m\right] dt = \frac{1}{\lambda} \sum_{i=1}^m \frac{1}{i} = \frac{1}{(0.0008)} \left(1 + \frac{1}{2}\right) = 1875 \text{ h}$$
(4.11)

Using the specified data values of parallel system reliability equation in Table 4.6 yields

$$R_{ps}(150) = \left[1 - \left\{1 - e^{-(0.0008)(150)}\right\}^2\right] = 0.9872 \quad (4.12)$$

Thus, the system mean time to failure and reliability are 1875 h and 0.9872, respectively.

4.5 Robust Design (or Taguchi Methods)

Robust design first developed by Taguchi is a powerful technique for improving reliability at low cost in a short time. Robust design is a statistical engineering methodology for optimizing product conditions so that product performance is minimally sensitive to various noise sources of variation (See Fig. 4.17). Since 1980s, it has been applied extensively to improve the quality of countless products and processes.

Robustness is defined as the ability of a product to perform its intended function consistently at the presence of noise factors such as environmental loads. Here, the

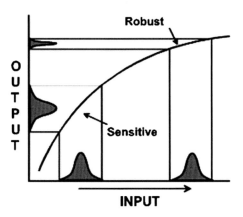

Fig. 4.17 Robust design—inputs that make the outputs less sensitive

noise factors are the variables that have adverse effects on the intended function and are impractical to control. Environmental stresses (or loads) are the typical noise factors. This definition is widely applied in the field of quality engineering to address initial robustness when the product service time is zero. If customer satisfaction over time is concerned, the effect of time should be taken into account.

Reliability of mechanical/civil system can be perceived as robustness over time or environmental conditions. A reliable product has a high robustness value under different use conditions. To achieve high robustness, Taguchi methods recommend the optimal design parameters insensitive to noise parameters.

Taguchi methods are originally a kind of method to improve the product quality and recently applied to engineering as robust design method. Professional statisticians have welcomed the goals and improvements brought about by Taguchi methods, particularly by Taguchi's development of designs for studying variation, but have criticized the inefficiency of some of Taguchi's proposals. As alternative methods, parametric Accelerated Life Testing in Chap. 7 will be studied.

4.5.1 A Specific Loss Function

To estimate these hidden quality costs, Taguchi's quality loss function (QLF) has been proposed. Taguchi's approach is different than the traditional approach of quality costs. In the traditional approach, if you have two products that one is within the specified limits and the other is just outside of the specified limits, the difference is small. Although the difference is small, the product within the limits is considered a good product. On the other hands, the outside one is considered a bad product. Taguchi disagrees with this approach. Taguchi believes that when a product moves from its target value, that move causes a loss. It does not matter if the move falls inside or outside the specified limits. For this reason, Taguchi developed the QLF to measure the loss that is associated with hidden quality costs. This loss happens when a variation causes the product to move away from its target value.

As seen in Fig. 4.18, QLF is a "U"-shaped parabola. The horizontal axis is tangent with the parabola at the target value. This is a quadratic loss function because it assumes that when a product is at its target value (T), the loss is zero.

Quality characteristics can be categorized into three situations: (1) On-target, minimum-variation, (2) Smaller-the-better, and (3) Larger-the-better.

4.5.1.1 On-Target, Minimum-Variation (for Example, a Mating Part in an Assembly)

In engineering design we frequently encounter the on-target characteristics. Due to production process variation, the characteristics are allowed to vary within a range,

4.5 Robust Design (or Taguchi Methods)

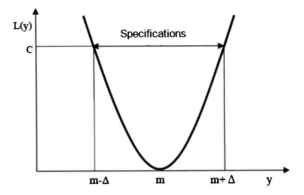

Fig. 4.18 Taguchi quality loss function for a nominal-the-best (on-target) characteristics

say $\pm\Delta$, where Δ is called the tolerance. Equation (4.13) describes the quality loss of this type of characteristics.

The unit loss is determined by the formula:

$$L(t) = k(y - T)^2 \qquad (4.13)$$

where k = a proportionality constant dependent upon the organization's failure cost structure, y = actual value of quality characteristic, T = target value of quality characteristic.

The value of k must first be determined before the loss can be estimated. To determine the value of k:

$$k = c/\Delta^2 \qquad (4.14)$$

where c = loss associated with the specification limit, and Δ − deviation of the specification from the target value.

The target function is described as

$$f(y) = 1/(y - t)^2 \qquad (4.15)$$

4.5.1.2 Smaller the Better—Variance (for Example, Carbon Dioxide Emissions)

If y is a smaller-the-better characteristic, its range can be written as $[0, d]$, where 0 is the target value and d is the upper specification limit. The quality loss function is obtained by substituting $T = 0$ into Eq. (4.13) and can be written as

$$L(t) = ky^2 \tag{4.16}$$

The target function is described as

$$f(y) = 1/y^2 \tag{4.17}$$

4.5.1.3 Larger the Better—Performance (for Example, Agricultural Yield)

If y is a larger-the-better characteristic, its range is $[d, \infty]$, where d is the lower limit. Because the reciprocal of a larger-the-better characteristic has the same quantitative behavior as a smaller-the-better one, the quality loss function can be obtained by substituting $1/y$ for y in Eq. (4.13). Then we have

$$L(t) = k\left(\frac{1}{y}\right)^2 \tag{4.18}$$

The target function is described as

$$f(y) = y^2 \tag{4.19}$$

Example 4.6 A product with on-target and minimum-variation has 100 (target value). The unit loss is determined by the formula:

$$L(t) = 40(y - 100)^2 \tag{4.20}$$

Find out the expectation of quality loss function of process line1 and line2
So expectation of quality loss function can be expressed as

$$E[L] = E\left[k(y-m)^2\right] = kVar[y] + k(E[y] - m)^2 \tag{4.21}$$

If process line1 has mean 96 and standard deviation 3, the expectation of quality loss function of process line1 is

$$E[L] = 40(3)^2 + 40(96 - 100)^2 = 1000\ \$ \tag{4.22}$$

If process line 2 has mean 98 and standard deviation 5, the expectation of quality loss function of process line2 is

4.5 Robust Design (or Taguchi Methods)

$$E[L] = 40(5)^2 + 40(98-100)^2 = 1160\ \$ \qquad (4.23)$$

If the standard variation of process line 2 decreases from 5 to 3, the cost reduction is

$$\Delta E[L] = 40(5)^2 - 40(3)^2 = 640\ \$ \qquad (4.24)$$

4.5.2 Robust Design Process

As seen in Fig. 4.19, robust design is a statistical engineering methodology for minimizing the performance variation of a product by choosing the optimal design conditions of the product to make the performance insensitive to noise factors. Taguchi realized that the best opportunity to eliminate variation is during the design of a product and its manufacturing process. Consequently, he developed a strategy for quality engineering that the process consists of three stages—system design, parameter design, and tolerance design.

4.5.2.1 System Design

System design involves selection of technology and components for use, design of system architecture, development of a prototype that meets customer requirement, and determination of manufacturing process. System design has significant impacts on cost, yield, reliability, maintainability, and many other performances of a product. It also plays a critical role in reducing product sensitivity to noise factors. If a system design is defective, the subsequent parameter design and tolerance design aimed at robustness improvement are ineffective. In recent years, some system design methodologies have emerged and shown effective, such as TRIZ

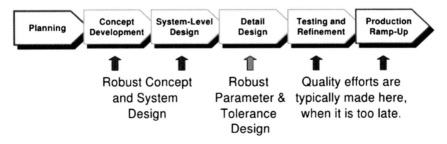

Fig. 4.19 Steps of robust design

(a problem-solving, analysis and forecasting tool derived from the study of patterns of invention in the global patent literature).

This step is indeed the conceptual design level, involving creativity, and innovation.

- Getting into the 'design space'
- Creating a feasible design
- Involves innovation

4.5.3 Parameter (Measure) Design

Parameter design aims at minimizing the sensitivity of the product performance to noise factors by setting its design parameters at the optimal levels. In this step, designed experiments are usually conducted to investigate the relationships between the design parameters and performance characteristics of the product. Using such relationships, one can determine the optimal setting of the design parameters.

Once the concept is established, the nominal values of the various dimensions and design parameters in the product need to be set, the detail design phase of conventional engineering. In many circumstances, this allows the parameters to be chosen so as to minimize the effects on performance arising from variation in environmental noise—loads. Strictly speaking, parameter design might signify the robust design

- Optimizing within the 'design space' (not changing anything fundamentally)
- Settings for the factors identified in systems design

4.5.4 Tolerance Design

With a successfully completed parameter design, tolerance design is to choose the tolerance of important parts to reduce the performance sensitivity to noise factors under cost constraints. Tolerance design may be conducted after the parameter design is completed. If the parameter design cannot achieve sufficient robustness, tolerance design is completed. In this step, the important parts whose variability has the largest effects on the product sensitivity are identified through experimentation. Then the tolerance of these parts is tightened by using higher grade parts based on the trade-off between the increased cost and the reduction in performance variability.

- Tightening tolerances on important factors
- Better materials/new equipment may be needed
- Has significant cost implications

4.5 Robust Design (or Taguchi Methods)

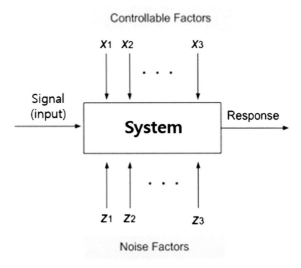

Fig. 4.20 Generic parameter diagram

4.5.5 A Parameter Diagram (P-Diagram)

A P-diagram in the mechanical/civil system that illustrates the input (signals), outputs (intended functions or responses), control factors, and noise factors. Figure 4.20 shows a generic P-diagram.

Signals are inputs from subsystems or modules to the system. The system transforms the signals into functional responses. Signals are essential to fulfilling the function of a system. Noise factors are variables that have adverse effects on robustness. Typical examples in the mechanical/civil system are a variety of random loads.

Control factors are the design parameters whose levels are specified by designers. The purpose of a robust design is to choose optimal levels of the parameters. In practice, mechanical/civil systems have a large of design parameters, which are not important in terms of the robustness. Thus, the key design parameters are included in a robust design. These key design parameters are identified by using engineering judgment, analytical study, a preliminary test, or field data (See Fig. 4.21).

4.5.6 Taguchi's Design of Experiment (DOE)

Design of experiment is a statistical technique for studying the effects of multiple factors on the experimental response. The factors are laid out in a structured array in which each row combination are conducted and response data are collected.

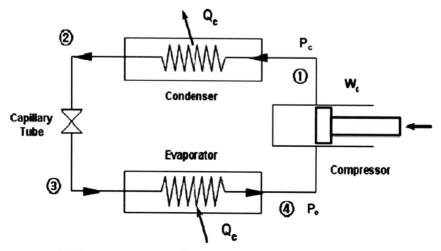

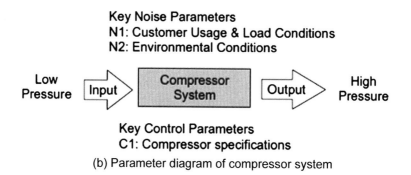

Fig. 4.21 Schematic diagram for mechanical system (example: compressor system)

Through experimental data analysis, we can choose the optimal levels of controls factors that minimize the sensitivity of the response to noise factors.

4.5.6.1 Orthogonal Arrays

An orthogonal array is a balanced fractional matrix in which each row represents the levels of factors of each run and each column represents the levels of specific factor that can be changed from each run. In a balanced matrix,

- All possible combinations of any two columns of the matrix occur equal number of times within the two columns. The two columns are also said to be orthogonal.

4.5 Robust Design (or Taguchi Methods)

Table 4.7 Experimental data

	A	B	AB	C	D	E	E	Experiment					S/N_i
	1	2	3	4	5	6	7	1	2	3	4	5	
1	0	0	0	0	0	0	0	34	22	29	14	25	−28.20
2	0	0	0	1	1	1	1	32	24	26	16	28	−28.22
3	0	1	1	0	0	1	1	24	18	25	27	22	−27.38
4	0	1	1	1	1	0	0	27	22	26	23	25	−27.84
5	0	0	1	0	1	0	1	30	25	27	29	20	−28.44
6	0	0	1	1	0	1	0	19	16	33	34	19	−28.04
7	0	1	0	0	1	1	0	25	33	24	25	21	−28.27
8	0	1	0	1	0	0	1	26	27	27	28	26	−28.57

- Each level of specific factor within a column has an equal number of occurrences within the column.

Example 4.7 If product on-target and minimum-variation have the following experimental data, find out the optimal factor combinations (Table 4.7) (Fig. 4.22).

First of all, make the simplified analysis for level 1 and 2 per A, B. C, D like Table 4.8.

Draw the Pareto chart and find the factors that consists of 80–90%, based on the accumulated total sum.

Then we know that A, B, and A × B occupy approximately 80% from Pareto charts. Find out the proper levels of A, B factors by Table 4.9.

Consequentially, we know that S/N ratio for A_0 and B_1 is the smallest among the levels of A, B, and A × B

4.5.7 Inefficiencies of Taguchi's Designs

Taguchi's robust design method uses parameter design to place the design in a position where random "noise" does not cause failure and to determine the proper design parameters and their levels. However, for a simple mechanical structure, a lot of design parameters should be considered in the Taguchi method's robust design process. Those mechanical/civil products with the missing or improper minor design parameters may result in recalls and loss of brand name value.

Interactions also are part of the real world. In Taguchi's arrays, interactions are confounded and difficult to resolve. Response surface methodology (RSM) is a "follow-up design" that resolves only the confounded interactions. RSM design may be used to explore possible high-order univariate effects of the remaining variables and have great statistical efficiency. The sequential designs of response surface methodology decreases experimental runs than would a sequence of Taguchi's designs. But Taguchi's designs still require a lot of experimental runs.

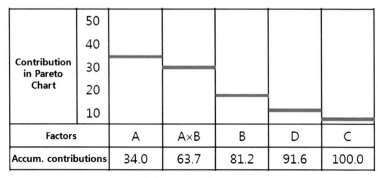

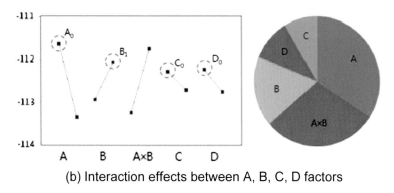

(b) Interaction effects between A, B, C, D factors

Fig. 4.22 Simplified analysis for A, B, C, D factors

Table 4.8 Simplified analysis for experimental data

Factor		A	B	AB	C	D	Total
S/N Ratio	Level 1	−111.64	−112.95	−113.26	−112.29	−112.24	−225.01
	Level 2	−113.37	−112.06	−113.26	−112.29	−112.24	
Level range		1.73	0.89	1.51	0.43	0.53	5.09
Contribution (%)		34.0	17.5	29.7	8.4	10.4	100.0

4.6 Reliability Testing

4.6.1 Introduction

Product lifetime can be estimated by using quantitative reliability testing methods. Reliability testing is performed to see whether intended function of product is feasible. Regardless of demonstrating successful requirement achievement, it causes

4.6 Reliability Testing

product failures to use concepts of load and strength under severe test conditions. It will be used to determine whether the product is adequate to meet the requirements of performance and reliability. Reliability testing during design and development therefore is mandatory to prove whether the lifetime of product is sufficient for customer requirements.

Product is the operational certainly for a stated time interval (or lifetime). The goal of product reliability is to develop product with a longer lifetime, based on the reliability target of product (or module). Today, the reliability of an element or of a system is defined as the probability that an item will perform its required function under given conditions for a stated time interval. The definition terms have to be explained as:

- Perform means that the item does not fail
- The given conditions include total physical environment, i.e., mechanical, electrical, and thermal conditions.
- The stated time interval can be very long (20 years, for telecommunication equipment), long (a few years), or short (a few hours or weeks, for space research equipment). But the time could be replaced by other parameters, such as: the mileage (of an automobile) or the number of cycles (i.e., for a relay unit).

Based on the observed data of reliability testing, maximum likelihood estimation (MLE) is a popular method to predict the product lifetime. MLE is the statistical method of estimating the parameters of a statistical model—some unknown mean and variance that are given to a data set. Maximum likelihood selects the set of values of the model parameters that maximizes the likelihood function.

For example, one may be interested in the lifetime of product (or module), but be unable to measure the lifetime of every single product in a population due to cost or time constraints. Assuming that the lifetime are normally distributed with some unknown mean and variance, the mean and variance can be estimated with MLE while only knowing the lifetime of some sample of the overall population. MLE would accomplish this by taking the mean and variance as parameters and finding particular parametric values that make the observed results the most probable given the model.

4.6.2 Maximum Likelihood Estimation

For a fixed set of data and underlying statistical model, the method of maximum likelihood estimates the set of values of the model parameters that maximizes the likelihood function. Intuitively, this maximizes the "agreement" of the selected model with the observed data, and for discrete random variables it indeed maximizes the probability of the observed data under the resulting distribution. MLE would give a unified approach to estimation, if the case of the normal distribution or many other problems is well defined.

One very good statistical method for the determination of unknown parameters of a distraction is the Maximum Likelihood Method. It assumes that the histogram of the failure frequency depicts the number of failure per interval. For larger test sample sizes n it is possible to derive function out of the histogram and thus to exchange the frequencies to the probabilities.

In this way it is possible to state, for example that during the first interval probably 3% of all failures will occur. In the second interval it is most likely that 45% of the failures occur, etc. According to theory, the probability L of test sample can be obtained by the product of the probability of the individual intervals.

Suppose there is a sample $t_1, t_2, \ldots, t_n$ of n independent and identically distributed failure times, coming from a distribution with an unknown probability density function $f_0(\cdot)$. On the other hand, supposed that the function f_0 belongs to a certain family of distributions $\{f(\cdot|\theta), \theta \in \Theta\}$ (where θ is a vector of parameters), called the parametric model, so that $f_0 = f(\cdot|\theta_0)$. The unknown value θ_0 is expected to as the true value of the parameter vector. An estimator $\hat{\theta}$ would be fairly close to the true value θ_0. The observed variables t_i and the parameter θ are vector components.

$$L(\Theta; t_1, t_2, \ldots, t_n) = f(t_1, t_2, \ldots, t_n | \Theta) = \prod_{i=1}^{n} f(t_i | \Theta) \qquad (4.25)$$

This function is called the likelihood. The idea of this procedure is to find a function f, for which the product L is maximized. Here, the function must possess high values of the density function f in the corresponding region with several failure times' t_i. At the same time only low value of f in regions with few failures are found. Thus, the actual failure behavior is accurately represented. If determined in this way, the function f gives the best probability to describe the test samples.

It is often more effective to use the log-likelihood function. Thus the product equation becomes an addition equation, which greatly simplifies the differentiation. Since the natural log is a monotonic function, this step is mathematically logical.

$$\ln L(\Theta; t_1, t_2, \ldots, t_n) = \sum_{i=1}^{n} \ln f(t_i | \Theta) \qquad (4.26)$$

By differentiating Eq. (4.26), the maximum of the log-likelihood function and thus the statistically optimal parameters Θ_l can be obtained as

$$\frac{\partial \ln(L)}{\partial \theta_l} = \sum_{i=1}^{n} \frac{1}{f(t_i; \Theta)} \cdot \frac{\partial f(t_i; \Theta)}{\partial \theta_l} = 0 \qquad (4.27)$$

These equations can be nonlinear in the parameters; therefore it is often useful to apply approximate numerical procedures. By the Likelihood function value the opportunity is given to estimate the quality of the adaptation of a distribution to the failure data. The greater the likelihood function value is, the better the conclusive distribution function represents the actual failure behavior. However, based on MLE, the characteristic life η_{MLE} from the reliability testing (or lifetime testing) can be estimated on the Weibull chart (Refer to Chap. 3.4).

4.6.3 Time-to-Failure Models

The failure time T of a product is a random variable. Time can take on different meanings depending on operational time, distance driven by a vehicle, and number of cycles for a periodically operated system. Time-to-Failure model usually provides all the tools for reliability testing, especially accelerated life testing data analysis. It is designed for use with complete (time-to-failure), right censored (suspended), interval or left censored data. Data can be entered individually or in groups.

Time-to-Failure Model has the following types:

- Arrhenius: a single stress model typically used when temperature is the accelerated stress.
- Inverse Power Law (IPL): a single stress model typically used with a nonthermal stress, such as vibration, voltage or temperature cycling.
- Eyring: a single stress model typically used when temperature or humidity is the accelerated stress.
- Temperature-Humidity: a double-Arrhenius model that is typically used when temperature and humidity are the acceleration variables.
- Temperature-Nonthermal: a combination of the Arrhenius and IPL relationships that is typically used when one stress is temperature and the other is nonthermal (e.g., voltage).

4.6.3.1 Arrhenius Equation

The Arrhenius equation proposed by Arrhenius in 1889 is a formula for the temperature dependence of reaction rates.

As seen in Fig. 4.23, reactivity modeling consists of computing the energy of the products, the reactants, and the transition state (TS) connecting them. These three points are the critical features on a reaction pathway. The difference between the energies of the transition state and reactants ($\Delta E_a = E_{TS} - E_r$) is the activation energy ΔE_a. The activation energy is important in understanding the rate of chemical reactions as expressed in the Arrhenius Equation which relates the rate constant K of a chemical reaction to its activation energy.

Fig. 4.23 Arrhenius equation

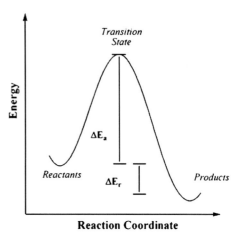

One of the earliest and most successful acceleration models predicts how time-to-fail varies with temperature. The Arrhenius equation empirically based model is known as

$$TF = A \exp\left(\frac{E_a}{kT}\right) \quad (4.28)$$

where T denoting temperature measured in degrees Kelvin at the point when the failure process takes place, k is Boltzmann's constant (8.617×10^{-5} in eV/K), and constant A is a scaling factor that drops out when calculating acceleration factors, with E_a denoting the activation energy, which is the critical parameter in the model. If Eq. (4.28) takes logarithm, the simple straight line can be obtained (Fig. 4.24).

The acceleration factor between a higher temperature T_2 and a lower temperature T_1 is given by

$$AF = \exp\left(\frac{E_a}{k}\left(\frac{1}{T_1} - \frac{1}{T_2}\right)\right) \quad (4.29)$$

Fig. 4.24 Arrhenius model

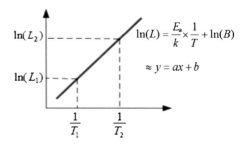

4.6 Reliability Testing

The value of E_a depends on the failure mechanism and the materials involved. It typically ranges from 0.3 or 0.4 up to 1.5, or even higher.

4.6.3.2 Inverse Power Law

In statistics, a power law is a functional relationship between two quantities, where a relative change in one quantity results in a proportional relative change in the other quantity, independent of the initial size of those quantities: one quantity varies as a power of another.

For example, the "life" of a product will go down as stress goes up. While this is not a hard and fast rule, very few systems do not behave in this intuitive fashion. This allows for shorter test times at higher levels of stress. With solid knowledge of the life–stress relationship, effective predictions of life at normal or usage conditions can be made.

The most important and widely used model for mechanical systems is the inverse power law (IPL). It has forms

$$TF = AS^{-n} \tag{4.30}$$

$$K = BS^n \tag{4.31}$$

The most critical factor is n, the life-stressor slope with s being stress applied to the system. A is a constant; in reality it relates the basic mechanical strength of the design to resist the stress applied to it. If Eq. (4.30) takes logarithm, the simple straight line can be obtained (Fig. 4.25).

4.6.3.3 Eyring Equation

For chemical reaction rate theory, Eyring acceleration model has led to a very general and powerful one. This model has several key features:

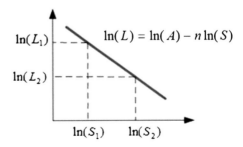

Fig. 4.25 Inverse power law

- It has a theoretical basis from chemistry and quantum mechanics.
- If a chemical process (chemical reaction, diffusion, corrosion, migration, etc.) is causing degradation leading to failure, the Eyring model describes how the rate of degradation varies with stress or, equivalently, how time to failure varies with stress.
- The model includes temperature and can be expanded to include other relevant stresses.
- The temperature term by itself is very similar to the Arrhenius empirical model, explaining why that model has been so successful in establishing the connection between the ΔE parameter and the quantum theory concept of "activation energy needed to cross an energy barrier and initiate a reaction."

The model for temperature and one additional stress takes the generic form:

$$TF = B \exp\left(\frac{E_a}{kT}\right) \times S^{-n} \quad (4.32)$$

where k is the Boltzmann constant, T is the thermodynamic temperature, and E_a is the activation energy.

The acceleration factor between a higher temperature T_2 and a lower temperature T_1 is given by

$$AF = \left(\frac{S_2}{S_1}\right)^n \exp\left(\frac{E_a}{k}\left(\frac{1}{T_1} - \frac{1}{T_2}\right)\right) \quad (4.33)$$

We know that the acceleration factor in Eyring equation Eq. (4.32) is similar to Eq. (7.17) that is derived from the generalized stress model Eq. (7.16).

4.6.4 Reliability Testing

The time and effort in testing can be significantly reduced by censored tests, and they can estimate the product lifetime. If a test trial is interrupted before all n test units have failed, a censored test may produce. If the interruption occurs after a given time, one is dealing with censoring of type I.

On the other hand, if a trial is interrupted after a given amount of test units r has failed, one is dealing with censoring of type II. The trials stop after 4 failures. The point in time at which the failure r occurs is a random variable. Thus, leaving the entire trial time length opens until the end of the trial.

4.6 Reliability Testing

The fact that n-r test units have not failed is taken into account by substituting r for n in the denominator of the approximation equation. With type I or II censoring it is necessary to estimate the characteristic lifetime η in the Weibull chart by extrapolating the best fit line beyond the last failure time. This is generally problematic as long as further failure mechanisms cannot be neglected. A statistical statement concerning the failure behavior can be obtained on the observed lifetime.

The procedures and methods for the assessment of complete data or censored data can be found in Table 4.9 and Fig. 4.26 (Table 4.10).

Table 4.9 Simplified analysis for A, B factors

	A_0	A_1
B_0	$S/N_{00} = -56.42$	$S/N_{10} = -56.53$
B_1	$S/N_{01} = -55.22$	$S/N_{11} = -56.84$

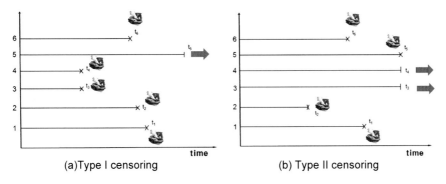

(a) Type I censoring   (b) Type II censoring

Fig. 4.26 Schematic of type I **a** and type II censoring **b**

Table 4.10 Overview of procedures for the assessment of censored data

Data type	Type of censor	Description	Procedure
Complete data $r = n$	No censoring	All samples have failed	Median procedure $F(t_i) \approx \frac{i-0.3}{n+0.4}$ For $i = 1, 2, \ldots, n$
Censored data $r < n$	Censoring Type I or Type II	Lifetime characteristics of all intact units are larger than the lifetime characteristics of the units r which failed last	Median procedure $F(t_i) \approx \frac{i-0.3}{n+0.4}$ For $i = 1, 2, \ldots, r$

Example 4.8 Select ten samples from a Integrated Circuit chip manufactured in august 2016 and perform the reliability testing at 120, 135, and 150 °C. Under 30 ° C normal conditions, search B10 life for 30° Con Weibull or excel program.

	Temperature, °C		
	120	130	150
1	3450	3300	2650
2	4340	3720	3100
3	4760	4180	3400
4	5320	4560	3800
5	5740	4920	4100
6	6160	5280	4400
7	6580	5640	4700
8	7140	6230	5100
9	8100	6840	5700
10	8960	7380	6400

Case (I) for 120 °C, as temperature data is complete with no censoring, plot them on Weibull chart. We can approximate sketch the best fit straight line through the entered points and determine the Weibull parameters $\hat{\beta} = 3.812$. At the $Q(t) = 63.2\%$ ordinate point, draw a straight horizontal line until this line intersects the fitted straight line. Draw a vertical line through this intersection until it crosses the abscissa. The value at the intersection of the abscissa is the estimate of $\hat{\eta} = 6692$.

i	120 °C	F(t) × 100
1	3450	6.73077
2	4340	16.3462
3	4760	25.9615
4	5320	35.5769
5	5740	45.1923
6	6160	54.8077
7	6580	64.4231
8	7140	74.0385
9	8100	83.6538
10	8960	93.2692

Case (II) Using Excel program, for 120 °C, we obtain the estimated shape parameter $\hat{\beta}$ = slope = 3.812, estimated characteristic life $\hat{\eta}$ ($F(t)$ is 63.2% ordinate point) = $e^{\frac{33.576}{3.812}}$ = 6692 hours.

4.6 Reliability Testing

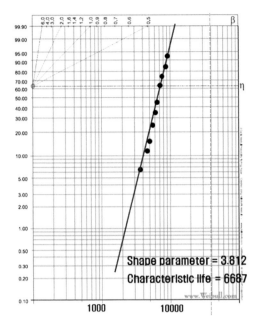

where $\ln[-\ln(1-0.63)] = -0.00576$, $-0.00576 = 3.811x - 33.57$

X: LN(t)	Y: LN(LN(1/(1-F(t))))
8.15	−2.66384
8.38	−1.72326
8.47	−1.20202
8.58	−0.82167
8.66	−0.50860
8.73	−0.23037
8.79	0.03293
8.87	0.29903
9.00	0.59398
9.10	0.99269

In the same way, we can obtain the analysis result of data for 130 and 150 °C. Plot them on the Weibull chart.

B10 life for 30 °C can be obtained from $L_{BX}^{\beta} = \left(\ln \frac{1}{1-x}\right) \cdot \eta^{\beta}$

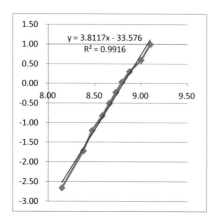

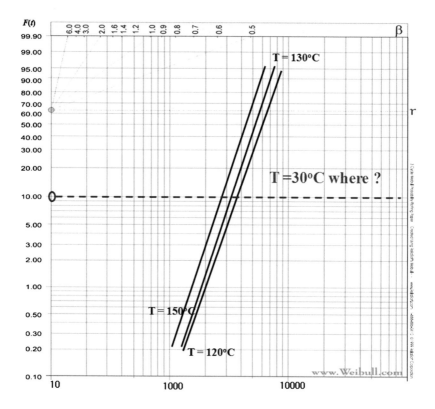

4.6 Reliability Testing

Temp, °C	B10 Life		1/T	Ln(B10)
120	3706		0.002544	8.217708
130	3355		0.00248	8.118207
150	2728		0.002363	7.911324

$$\ln(L) = \ln(A) + \frac{E_a}{k} \times \frac{1}{T} \qquad \Longleftrightarrow \qquad y = 3.8782 + 1707.4x$$

$$\ln(A) = 3.878204, \quad \frac{E}{k} = 1707.392$$

$A = \exp(a) = 48.3373$, $E = 1707.392 \times (8.617 \times 10^{-5}) = 0.147126$ eV
So the estimated life-stress model from Eq. (4.28) is

$$TF = A \exp\left(\frac{E_a}{kT}\right) = 48.3373 \exp\left(\frac{1707.392}{T}\right)$$

So B10 life for 30 °C is obtained from

$$TF = 48.3373 \exp\left(\frac{1707.392}{30 + 273.16}\right) = 13,506$$

Chapter 5
Load Analysis

Abstract This chapter will explain how to model the mechanical/civil systems—automobiles, aircraft, satellites, rockets, space stations, ships, bridge, and building subjected to the random loading. Products have their own particular structural loads in the field. A typical pattern of repeated load or overloading may cause structural failure in product lifetime. Such possibility should be assessed in the design phase whether structure subjected to loads endures in its lifetime. Modeling is a mathematical representation of the dynamics system to describe the real world used by traditional system modeling method like Newtonian. Here, as alternative method, the bond graph will be introduced because it is easily applicable to the mechanical/civil systems. If products are modeled, the time response of system simulation for (random) dynamic loads will obtain. As the time response is simplified and counted as a sinusoidal input, the rain-flow counting method and miner's rule can assess the system damage. Because there are a lot of assumptions, this analytic methodology is exact but complex to reproduce the reliability disasters due to the design failures. So, we should develop the final solutions—experimental method like parametric ALT that will be discussed in Chap. 7. Load analysis will be helpful to figure out the failure of problematic parts and finally discover them in the reliability-embedded design process.

Keywords Load analysis · Mathematical modeling · Bond graph · Miner's rule · Rain-flow counting

5.1 Introduction

Loads cause stresses, deformations, and displacements in the structures of product. Assessment of their effects can be implemented by the structural modeling and its analysis using finite element. In a result, repeated load or overloading may cause structural failure.

Two generic types of mechanical static or dynamic loading exist. A static load—tension or compression can exhibit motion or permenent change like dislocation if repeated in a lifetime. Eventually, they will be a permanent deformation. The examples of static loading are as following:

- Structural load and deflection versus material stress and strain
- Tension and compression loads
- Torsion and bending loads

A dynamic load, sometimes also referred to as probabilistic loads, is a force exerted by a moving body on a resisting member, usually in a relatively short or long period of time. Because such loads are usually unstable, we can say the dynamic load. Dynamic loads involve motion and therefore are time varying load conditions. The examples of dynamic loading are as following:

- Impact, vibration and shock loads
- Unbalanced inertia loads

An impact load is one whose time of application on a material is less than one-third of the natural period of vibration of that material. A variety of cyclic loads on a structure can lead to fatigue damage, cumulative damage, or fracture. These loads come from repeated loadings on a structure or can be due to vibration.

5.2 Modeling of Mechanical System

5.2.1 Introduction

The modeling of mechanical product is a mathematical representation of the dynamics structural systems to figure out their physical characteristics. Typical modeling methods—Newton, Lagrange, Hamiltonian mechanics, and D'Alembert's Law—are commonly used in dynamic system. As an output, models might describe the system behavior that can be represented in random variables (or state space). In a result, the state space is expressed as vectors and provides a convenient and compact way to analyze systems with multiple inputs and outputs.

When observed in most mechanical/civil components, loads in field follows a more or less random curve that constant load amplitudes are quite seldom. For example, the trajectories of automobiles possess completely random stochastic load curves due to the street roughness, car speed, and environmental conditions. Moreover, for airplane, a mean load change repetitively occurs on the wing of a transportation airplane when it takes off or lands (see Fig. 5.1).

On the other hands, the load of the gas turbine blade in an airplane is to a large extent deterministic that there is no randomness in the system states, though the load sequence is still variable. With simple algorithms and fast processors an online load measurement for parts can be directly measured during operation.

However, a measurement during operation is quite time-consuming and actually impossible to figure out the whole transmitted loads in product lifetime. To do that,

5.2 Modeling of Mechanical System

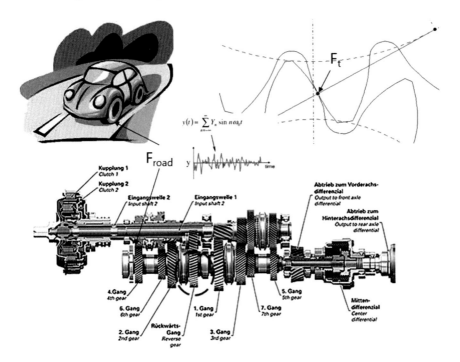

Fig. 5.1 Operational loads due to the random vibration on road

the engineer depends on the mathematical modeling, analysis, and its response, such as the Newtonian model that was deveeloped for long time ago (see Fig. 5.2).

5.2.2 D'Alembert's Modeling for Automobile

Engineer uses D'Alembert's principle and free body diagram to model mechanical system. If there is an automobile that is used for transportation, we can model a simple system with a mass that is separated from a wall by a spring and a dashpot. The mass could represent an automobile, with the spring and dashpot representing the automobile's bumper. If only horizontal motion and forces are considered, it is represented in Fig. 5.3.

The free body diagram is a drawing method showing all external forces acting on a body. There is only one position in this system defined by the variable "x" that is positive to the right. It is assumed that $x = 0$ when the spring is in its relaxed state. As seen in Fig. 5.4, there are four forces to develop a model from the free body diagram: (1) An external force (F_e) such as friction force and air resistance force, (2) A spring force that will be a force from the spring, $k \cdot x$, to the left, (3) A dashpot force that will be a force from the dashpot, $b \cdot v$, to the left, (4) Finally,

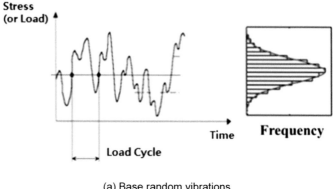

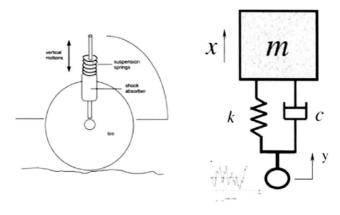

(b) A simplified modeling of the automobiles

Fig. 5.2 Random loads and modeling of the automobiles by Newtonian modeling

there is the inertial force which is defined to be opposed to the defined direction of motion. This is represented by $m \cdot a$ to the left.

Newton's second law states that an object accelerates in the direction of an applied force, and that this acceleration is inversely proportional to the force, or

$$\sum_{\text{all lexternal}} F = m \cdot a \tag{5.1}$$

Subtracting the right-hand side results in D'Alembert's principle,

$$\sum_{\text{all lexternal}} F - m \cdot a = 0 \tag{5.2}$$

5.2 Modeling of Mechanical System

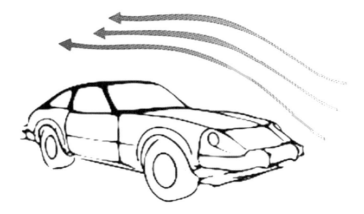

(a) Typical automobile subjected to wind flow

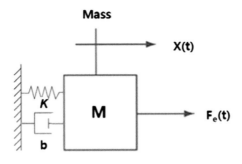

(b) Mass-spring-dashpot system

Fig. 5.3 Typical mechanical automobile modeling

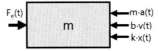

Fig. 5.4 Completed free body diagram for automobile modeling

If we consider the $m \cdot a$ term to be inertia force (or D'Alembert's force), D'Alembert's law will be left

$$\sum_{\text{all}} F \cdot \delta r = 0 \qquad (5.3)$$

To visualize this consider pushing against a mass (in the absence of friction) with your hand in the positive direction. Your hand experiences a force in the direction opposite to that of the direction of the force (this is the $-m \cdot a$ term). The inertial force is always in a direction opposite to the defined positive direction. We sum all of these forces to zero and get

$$F_e(t) - ma(t) - bv(t) - k \cdot x(t) = 0 \qquad (5.4)$$

In other words, we can change

$$m\frac{d^2x}{dt^2} + b\frac{dx}{dt} + kx(t) = F_e(t) \qquad (5.5)$$

5.3 Bond Graph Modeling

5.3.1 Introduction

Bond graph is an explicit graphical tool for modeling multidisciplinary dynamic systems including components from different engineering areas—the mechanical/civil, the electrical, the thermal, and the hydraulic system. When designing a new dynamic system, it is a good method to utilize a graphical representation for communicating other engineers to express the dynamic modeling. In engineering disciplines, linear graphs have long traditions among several graphical representation means.

In 1959, bond graph method was developed by Professor Henry Payner and his former students at MIT, who gave the revolutionary idea of portraying systems in terms of power bonds, connecting the elements of the physical system to the so-called junction structures, which were manifestations of the constraints. This power exchange portray of a system is called bond graph.

In 1961, the Paynter's books were published as entitled "Analysis and Simulation of Simulation of Multiport Systems." In 2006, the three authors have published the fourth edition entitled as "System Dynamics—Modeling and Simulation of Mechatronic Systems". Now several disciplines of bond graph have been widely accepted in the world as a modeling methodology. There are many literatures about bond graph method and its applications to analyze dynamic systems.

In a result, this method will give a brief description for analyzing loads applied to structure and understanding its work (Figs. 5.5 and 5.6).

5.3 Bond Graph Modeling

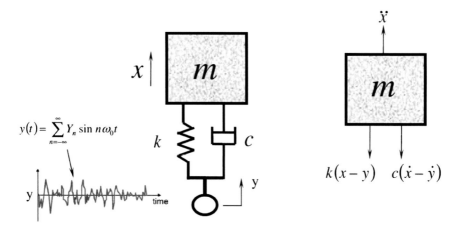

Fig. 5.5 A typical modeling of the automobiles subjected to repetitive random vibrations

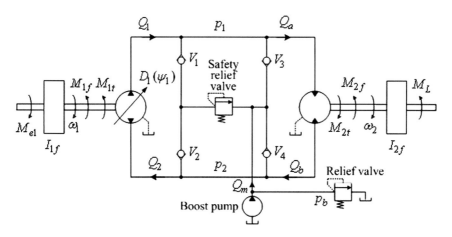

Fig. 5.6 Typical hydrostatic transmission modeling

5.3.2 Basic Elements, Energy Relations, and Causality of Bond Graph

A Bond Graph is a graphical representation of a physical dynamic system. It is similar to the block diagram and signal-flow graph. While, the symbols in bond graph represent bidirectional exchange of physical energy, those in block diagrams and signal-flow graphs represent unidirectional flow of information. Bond graph also can be applicable in multi-energy domain—mechanical/civil, electrical, and hydraulic system.

The dynamic systems analysis is relatively simple when the steady-state behavior or the few degrees of freedom has. In most of the cases, the main concern of engineers is to establish the mathematical model that represents the dynamic behavior of the system and how the different parameters influence the system behavior, because the system dynamic equations are usually partial differential equations, whose solutions require deep mathematical knowledge.

The fundamental bases of the bond graph theory, energy flow is a basic element in a system. It flows from one or more sources and is temporarily stored in system components or partially dissipated in resistances as heat, and finally arrives at "loads," where it produces some desired effects. Power is the rate of energy flow without direction.

Bond Graph represents this power flow between two systems. This flow is symbolized through an arrow (Bond) as Fig. 5.7 illustrated. Each bond represents the instantaneous energy flow or power. The flow in each bond is denoted by a pair of variables called 'power variables' whose product is the instantaneous power of the bond. Because power is not easy to measure directly, engineers can be represented as two temporary variables—flow and effort. Every domain has a pair of effort and flow variable. For example in mechanical system, flow represents the "velocity" and effort the "force," in electrical system, flow represents the "current" and effort the "voltage." The product of both temporary variables—power is represented as:

$$P = e(t) \cdot f(t). \tag{5.6}$$

The method makes possible the simulation of multiple physical domains, such as mechanical, electrical, thermal, hydraulic, etc. Flows and efforts should be identified with a particular variable for each specific physical domain, which is working. Table 5.1 also shows the physical meanings of the variables in different domains.

The Bond Graph is composed of the "bonds" which link together "1-port,"s "2-port," and "3-port" elements. Whether power in bond graph is continuous or not, every element is represented by a multi-port. Ports are connected by bonds. The basic blocs of standard bond graph theory are listed in Table 5.2.

For 1-ports, there are effort sources, flow sources, C-type elements, I-type Elements, and R-type Elements that can connect power discontinuously. For

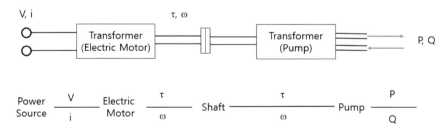

Fig. 5.7 Power flow in bond graph for electric-hydraulic system

5.3 Bond Graph Modeling

Table 5.1 Energy flow in the multi-port physical system

Modules	Effort, $e(t)$	Flow, $f(t)$
Mechanical translation	Force, $F(t)$	Velocity, $V(t)$
Mechanical rotation	Torque, $\tau(t)$	Angular velocity, $\omega(t)$
Compressor, pump	Pressure difference, $\Delta P(t)$	Volume flow rate, $Q(t)$
Electric	Voltage, $V(t)$	Current, $i(t)$
Thermal	Temperature, T	Entropy change rate, ds/dt
Chemical	Chemical potential, μ	Mole flow rate, dN/dt
Magnetic	Magneto-motive force, e_m	Magnetic flux, φ

Table 5.2 Basic elements of bond graph

Elements		Symbol	Relation equations
1-port elements (sources)	Effort	S_e—	$S_e = e(t)$
	Flow	S_f—	$S_f = f(t)$
1-port elements	C-type elements	C—	$\frac{de(t)}{dt} = \frac{1}{C}f(t)$
	I-type elements	I—	$\frac{df(t)}{dt} = \frac{1}{I}e(t)$
	R-type elements	R—	$e(t) = R \cdot f(t)$
2-port elements	Transformer	—1 TF 2—	$e_2(t) = TF \cdot e_1(t)$
	Gyrator	—1 GY 2—	$e_2(t) = GY \cdot f_1(t)$
3-port junction elements	0-junction	—1 0 2—	$e_2(t) = e_1(t)$
	1-junction	—1 1 2—	$f_2(t) = f_1(t)$

2-ports, there are Transformer and Gyrator that can connect power continuously. For 3-ports, there are 0-junction and 1-junction that can make up the network.

Power bonds may join at one of two kinds of junctions: a "0" junction and a "1" junction. In a "0" junction, the flow, and the efforts satisfy Eqs. (5.7) and (5.8):

$$\sum \text{flow}_{\text{input}} = \sum \text{flow}_{\text{output}} \qquad (5.7)$$

$$\text{effort}_1 = \text{effort}_2 = \cdots = \text{effort}_n \qquad (5.8)$$

This corresponds to a node in an electrical circuit (where Kirchhoff's current law applies). In a "1" junction, the flow and the efforts satisfy Eqs. (5.9) and (5.10):

$$\sum \text{effort}_{\text{input}} = \sum \text{effort}_{\text{output}} \qquad (5.9)$$

$$\text{flow}_1 = \text{flow}_2 = \cdots = \text{flow}_n \qquad (5.10)$$

This corresponds to force balance at a mass in a system. An example of a "1" junction is a resistor in series. In junction, the premise of energy conservation is assumed, no lost is allowed. There are two additional variables, important in the description of dynamic systems.

For any element with a bond with power variables—effort and flow, the energy variation from t_0 to t can be expressed by:

$$H(t) - H(t_0) = \int_{t_0}^{t} e(\tau) f(\tau) d\tau \qquad (5.11)$$

For C-type elements, e (effort) is a function of q (displacement). If displacement is differentiated, flow is obtained as

$$q(t) = \int f(t) dt \Rightarrow \frac{dq}{dt} = f(t) \qquad (5.12)$$

If Eq. (5.11) is changing variables from t to q, the linear case can be expressed as:

$$H(q) - H(q_0) = \frac{1}{2C} \left(q^2 - q_0^2 \right) \qquad (5.13)$$

For I-type elements, f (flow) is a function of p (momentum). If momentum is differentiated, effort is obtained as

$$p(t) = \int e(t) dt \Rightarrow \frac{dp}{dt} = e(t) \qquad (5.14)$$

If Eq. (5.11) is changing variables from t to p, the linear case can be expressed as:

$$H(p) - H(p_0) = \frac{1}{2I} \left(p^2 - p_0^2 \right) \qquad (5.15)$$

Resistor elements represent situations where energy dissipates—electrical resistor, mechanical damper, and coulomb frictions. In these sorts of elements, there is a relationship between flow and effort as the Eq. (5.16) shows. The value of "R" can be constant or function of any system parameter including time.

5.3 Bond Graph Modeling

$$e(t) = R \cdot f(t) \quad (5.16)$$

Compliance elements represent the situations where energy stores—electrical capacitors, mechanical springs, etc. In these sorts of elements, there is a relationship between effort and displacement variable as the Eq. (5.17) shows. The value of "K" can be constant or function of any system parameter including time.

$$e(t) = K \cdot q(t) \quad (5.17)$$

Inertia elements represent the relationship between the "flow" and Momentum (electrical coil, mass, moment of inertia, etc.) as the Eq. (5.18) shows. The value of "I" tends to be constant

$$p(t) = I \cdot f(t) \quad (5.18)$$

A transformer adds no power but transforms it, such as an electrical transformer or a lever. Transformers represent those physical phenomena that are variation of the values of output flow and effort on the values of input flow and effort. If the transformation ratio is given by the "TF" value, then the relationship between input and output is shown in Eqs. (5.19) and (5.20).

$$e_{output}(t) = TF \cdot e_{input}(t) \quad (5.19)$$

$$f_{output}(t) = \frac{1}{TF} \cdot f_{input}(t) \quad (5.20)$$

One is the "half-arrow" sign convention. This defines the assumed direction of positive energy flow. As with electrical circuit diagrams and free body diagrams, the choice of positive direction is arbitrary, with the caveat that the analyst must be consistent throughout with the chosen definition. The other feature is the "causal stroke." This is a vertical bar placed on only one end of the bond. It is not arbitrary (Fig. 5.8).

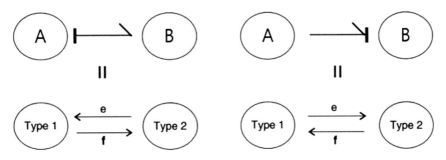

Fig. 5.8 "Half-arrow" sign convention and meaning of the causal stroke

On each Bond, one of the variables must be the cause and the other one the effect. This can be deduced by the relationship indicated by the arrow direction. Effort and flow causalities always act in opposite directions in a Bond. The causality assignment procedure chooses who sets what for each bond. Causality assignment is necessary to transform the bond graph into computable code.

Any port (single, double or multi) attached to the bond shall specify either "effort" or "flow" by its causal stroke, but not both. The port attached to the end of the bond with the "causal stroke" specifies the "flow" of the bond. And the bond imposes "effort" upon that port. Equivalently, the port on the end without the "causal stroke" imposes "effort" to the bond, while the bond imposes "flow" to that port.

Once the system is represented in the form of bond graph, the state equations that govern its behavior can be obtained directly as a first order differential equations in terms of generalized variables defined above, using simple and standardized procedures, regardless of the physical domain to which it belongs, even when interrelated across domains.

5.3.3 Case Study: Hydrostatic Transmission (HST) in Seaborne Winch

The winch structure is designed for launching, owing, and handling the cable and array in ship. The operation conditions of seaborne winch can be varied such as operation conditions—sea state, ship speed, and towing cable length. Because its operation requires high tension, seaborne winch is commonly used by the hydrostatic transmission (HST). It consists of electric motor, pump, piping, hydraulic motor, and loads. Tension and the response characteristics under the states of launching, towing, and hauling should be known before the design of HST. Tension data can be obtained from tension experiment. However, as an experiment, obtained the exact time response characteristics has many difficulties. And many previous design methods for HST involve extensive calculations because energy type of HST changes from mechanical to hydraulic, and then mechanical system. Bond Graph can easily model HST system and the dynamic response (Fig. 5.9).

HST as shown in Fig. 5.10 is commonly divided into electric motor, hydraulic pump, piping system, safety switches, and hydraulic motor. A rotating electric motor operates a hydraulic pump, which supplies oil to pipe system.

As cylinders in a hydraulic motor are filled with oil, shat rotates load. Therefore, HST is a kind of the closed-loop power transmission. The effort and flow in the rotating mechanical/civil system are torque and angular velocity, respectively. If two elements are integrated, they became momentum and volume. No matter what systems in HST may be, power does not change.

Bond Graph of electric motor and hydraulic pump is shown in Fig. 5.11. Source flow SF_{11} indicates an electric motor with constant angular velocity. It is assumed

5.3 Bond Graph Modeling

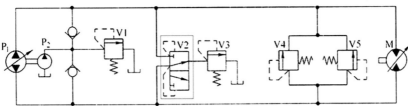

Fig. 5.9 Hydraulic-driven winch system in ship-borne

that a 10% among total torque perishes out by resistance element R_{12}. Transducer element MTF_{11} represents the capacity of a variable piston pump, which can control capacity with swivel angle α_o.

A bulk modulus B with implies oil compressibility chooses 10,000 bar among 6000–12,000 bar. Fluid condensers $C_{23} = C_{21}$ are described as V/B. Fluid inertia I_{24} represents oil mass. Using the least square method, resistance R_{22} and R_{26} are calculated from the pump and motor leakage. Because pipe flow is laminar, fluid resistance R_{25} can be calculated. Motor capacity TF_{3128} is determined from the number of filling cylinders. Moment of inertia of drum and flange I_{33} can be calculated. It is assumed that torque loss of flange R32 is about 10%. When bond graph is drawn from top and bottom—starting with the electric motor and ending with the load, a total bond graph and derivation of the state equations of a HST is represented as:

To obtain nondimensional state equations, nondimensional variables are introduced as

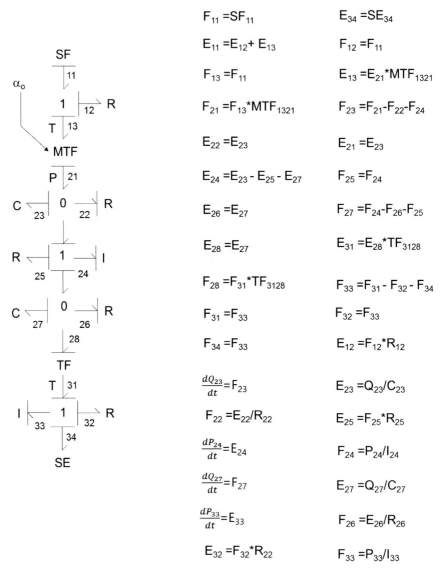

Fig. 5.10 Bond graph and derivation of the state equations of the Hydrostatic transmission in seaborne winch

$$\tilde{p} = \frac{P}{I\bar{Q} \text{ or } \omega}, \quad \tilde{q} = \frac{Q}{C\bar{P}}, \quad \tilde{t} = \frac{t}{\omega_n^{-1}} = \frac{1}{\sqrt{IC}} \tag{5.21}$$

$$\frac{dQ}{dt} = \frac{d(C\bar{P}\tilde{q})}{dt} = \frac{d(C\bar{P}\tilde{q})}{d\tilde{t}}\frac{d\tilde{t}}{dt} = C\bar{P}\omega_n\dot{\tilde{q}} \tag{5.22}$$

5.3 Bond Graph Modeling

Fig. 5.11 Electric motor and hydraulic pump modeling

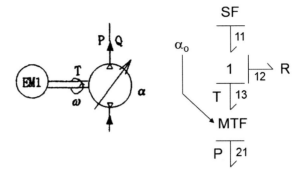

$$\frac{dP}{dt} = \frac{d(I\bar{Q}\tilde{p})}{dt} = \frac{d(I\bar{Q}\tilde{p})}{d\tilde{t}}\frac{d\tilde{t}}{dt} = I\bar{Q}\omega_n \dot{\tilde{p}} \quad (5.23)$$

where P, Q, and t are dimensional integral of pressure, volume, and time $\tilde{p}$, $\tilde{q}$ and $\tilde{t}$ are nondimensional integral of pressure, volume, and time, respectively. Therefore, nondimensional state equations are derived as:

$$\frac{C_{23}\omega_n \bar{P}}{SF_{11}MTF_{2113}} \dot{\tilde{q}}_{23} + \frac{\bar{P}}{R_{23}SF_{11}MTF_{2113}} \tilde{q}_{23} + \frac{\bar{Q}}{SF_{11}MTF_{2113}} \tilde{p}_{24} = 1 \quad (5.24)$$

$$\frac{I_{24}\omega_n}{R_{25}} \dot{\tilde{p}}_{24} - \frac{\bar{P}}{R_{25}\bar{Q}} \tilde{q}_{23} + \tilde{p}_{24} + \frac{\bar{P}}{R_{25}\bar{Q}} \tilde{q}_{27} = 0 \quad (5.25)$$

$$C_{27}\omega_n R_{26} \dot{\tilde{q}}_{27} - \frac{\bar{Q}R_{26}}{\bar{P}} \tilde{p}_{24} + \tilde{q}_{27} + \frac{\omega TF_{3128}R_{26}}{\bar{P}} \tilde{p}_{33} = 0 \quad (5.26)$$

$$\frac{I_{33}\omega\omega_n}{SE_{34}} \dot{\tilde{p}}_{33} - \frac{\bar{P}TF_{3128}}{SE_{34}} \tilde{q}_{27} + \frac{R_{32}\omega}{SE_{34}} \tilde{p}_{33} = 0 \quad (5.27)$$

To investigate the dynamic stability of the system, simple asymptotic approach can be used and perturbations around stable points are expressed as:

$$\tilde{q}_{23} = \tilde{q}_{230} + \varepsilon^1 \tilde{q}_{231} + O(\varepsilon^2) \quad (5.28)$$

$$\tilde{p}_{24} = \tilde{p}_{240} + \varepsilon^1 \tilde{p}_{241} + O(\varepsilon^2) \quad (5.29)$$

$$\tilde{q}_{27} = \tilde{q}_{270} + \varepsilon^1 \tilde{q}_{271} + O(\varepsilon^2) \quad (5.30)$$

$$\tilde{p}_{33} = \tilde{p}_{330} + \varepsilon^1 \tilde{p}_{331} + O(\varepsilon^2) \quad (5.31)$$

where ε' is very small value.

Substitute Eq. (5.24)–(5.27) into (5.28)–(5.31), then the terms of ε^0 is yield

$$\frac{\bar{P}}{R_{22}SF_{11}MTF_{2113}}(\tilde{q}_{230}) + \frac{\bar{Q}}{SF_{11}MTF_{2113}}\tilde{p}_{240} = 1 \tag{5.32}$$

$$-\frac{\bar{P}}{R_{25}\bar{Q}}\tilde{q}_{230} + \tilde{p}_{240} + \frac{\bar{P}}{R_{25}\bar{Q}}\tilde{q}_{270} = 0 \tag{5.33}$$

$$-\frac{\bar{Q}R_{26}}{\bar{P}}\tilde{p}_{240} + \tilde{q}_{270} + \frac{\omega TF_{3128}R_{26}}{\bar{P}}\tilde{p}_{330} = 0 \tag{5.34}$$

$$-\frac{\bar{P}TF_{3128}}{SE_{34}}\tilde{q}_{270} + \frac{R_{32}\omega}{SE_{34}}\tilde{p}_{330} = -1 \tag{5.35}$$

And then the terms of ε^1 is yield

$$\frac{C_{23}\omega_n \bar{P}}{SF_{11}MTF_{2113}}\dot{\tilde{q}}_{231} + \frac{\bar{P}}{R_{23}SF_{11}MTF_{2113}}\tilde{q}_{231} + \frac{\bar{Q}}{SF_{11}MTF_{2113}}\tilde{p}_{241} = 0 \tag{5.36}$$

$$\frac{I_{24}\omega_n}{R_{25}}\dot{\tilde{p}}_{241} - \frac{\bar{P}}{R_{25}\bar{Q}}\tilde{q}_{231} + \tilde{p}_{241} + \frac{\bar{P}}{R_{25}\bar{Q}}\tilde{q}_{271} = 0 \tag{5.37}$$

$$C_{27}\omega_n R_{26}\dot{\tilde{q}}_{271} - \frac{\bar{Q}R_{26}}{\bar{P}}\tilde{p}_{241} + \tilde{q}_{271} + \frac{\omega TF_{3128}R_{26}}{\bar{P}}\tilde{p}_{331} = 0 \tag{5.38}$$

$$\frac{I_{33}\omega\omega_n}{SE_{34}}\dot{\tilde{p}}_{331} - \frac{\bar{P}TF_{3128}}{SE_{34}}\tilde{q}_{271} + \frac{R_{32}\omega}{SE_{34}}\tilde{p}_{331} = 0 \tag{5.39}$$

If the perturbed Eqs. (5.36)–(5.39) are expressed as state–space form $[dx/dt] = [A][X]$, then

$$\begin{bmatrix} \dot{\tilde{q}}_{231} \\ \dot{\tilde{p}}_{241} \\ \dot{\tilde{q}}_{271} \\ \dot{\tilde{p}}_{331} \end{bmatrix} = \begin{bmatrix} -\frac{1}{R_{22}C_{23}\omega_n} & -\frac{\bar{Q}}{C_{23}\omega_n \bar{P}} & 0 & 0 \\ \frac{-\bar{P}}{\bar{Q}I_{24}\omega_n} & -\frac{R_{25}}{I_{24}\omega_n} & -\frac{\bar{P}}{\bar{Q}\omega_n I_{24}} & 0 \\ 0 & \frac{1}{\bar{P}C_{27}\omega_n} & -\frac{1}{C_{27}\omega_n R_{26}} & -\frac{\omega TF_{3128}}{\bar{P}C_{27}\omega_n} \\ 0 & 0 & \frac{-\bar{P}TF_{3128}}{I_{33}\omega\omega_n} & -\frac{R_{32}}{I_{33}\omega_n} \end{bmatrix} \begin{bmatrix} \tilde{q}_{231} \\ \tilde{p}_{241} \\ \tilde{q}_{271} \\ \tilde{p}_{331} \end{bmatrix} \tag{5.40}$$

To investigate the dynamic stability of the nondimensional state Eq. (5.40), eigenvalue of the bond graph can be represented as a state equation form $|A - \lambda I|[X] = 0$. The system is unstable if eigenvalue are $\lambda > 0$ and the system is stable $\lambda < 0$. When the state equations are represented as state space form of

5.3 Bond Graph Modeling

$$\begin{bmatrix} \frac{dQ_{23}}{dt} \\ \frac{dP_{24}}{dt} \\ \frac{dQ_{27}}{dt} \\ \frac{dP_{33}}{dt} \end{bmatrix} = \begin{bmatrix} -\frac{1}{C_{23}R_{22}} & -\frac{1}{I_{24}} & 0 & 0 \\ \frac{1}{C_{23}} & -\frac{R_{25}}{I_{24}} & -\frac{1}{C_{27}} & 0 \\ 0 & \frac{1}{I_{24}} & -\frac{1}{C_{27}R_{26}} & -\frac{TF_{3128}}{I_{33}} \\ 0 & 0 & \frac{TF_{3128}}{C_{27}} & -\frac{R_{32}}{I_{33}} \end{bmatrix} \begin{bmatrix} Q_{23} \\ P_{24} \\ Q_{27} \\ P_{33} \end{bmatrix} + \begin{bmatrix} MTF_{2113} \\ 0 \\ 0 \\ 0 \end{bmatrix} [SF_{11}] + \begin{bmatrix} 0 \\ 0 \\ 0 \\ -1 \end{bmatrix} [SE_{34}]$$

(5.41)

When Eq. (5.41) is integrated, the pump pressure and motor pressure are obtained as

$$\begin{bmatrix} \bar{P}_{pump} \\ \bar{P}_{motor} \end{bmatrix} = \begin{bmatrix} 1/C_{23} & 0 \\ 0 & 1/C_{27} \end{bmatrix} \begin{bmatrix} Q_{23} \\ Q_{27} \end{bmatrix}$$

(5.42)

HST simulations are classified as models of low speed, high, and maximum tension. The tension values might be obtained by the drag force analysis of cable.

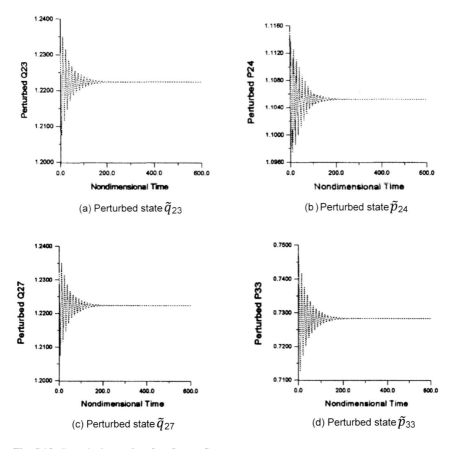

Fig. 5.12 Perturbed state $\tilde{q}_{23}$, $\tilde{p}_{24}$, $\tilde{q}_{27}$ and $\tilde{p}_{33}$

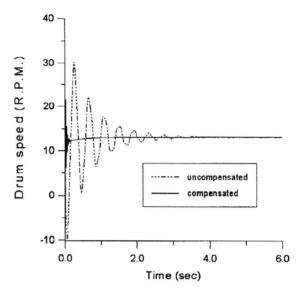

Fig. 5.13 Simulation results for hydrostatic transmission

A steady solution of ε^0 equation and eigenvalues from high speed, low speed, and maximum tension are calculated as stable. The values of (a) perturbed state Q_{23} (b) perturbed state P_{24} (c) perturbed state Q_{27} (d) perturbed state P_{33} from high speed mode are shown in Fig. 5.12. State variables are converged after they perturbed around steady-state value ε^0. It can figure out that simulations results with a big overshoot reach a stead-state value (Fig. 5.13).

5.3.4 Case Study: Failure Analysis and Redesign of a Helix Upper Dispenser

The mechanical icemaker system in a side-by-side (SBS) refrigerator with a dispenser system consists of many structural parts. Depending on the customer usage conditions, these parts receive a variety of mechanical loads in the ice making process. Ice making involves several mechanical processes: (1) the filtered water is pumped through a tap line supplying the tray; (2) the cold air in the heat exchanger chills the water tray; and (3) after ice is made, the cubes are harvested, stocking the bucket until it is full. When the customer pushes the lever by force, cubed or crushed ice is dispensed. In the United States, the customer typically requires an SBS refrigerator to produce 10 cubes per use and up to 200 cubes a day. Ice production may be influenced by uncontrollable customer usage conditions such as water pressure, ice consumption, refrigerator notch settings, and the number of times the door is opened. When the refrigerator is plugged in, the cubed ice mode is

5.3 Bond Graph Modeling

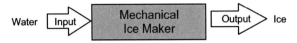

Fig. 5.14 Robust design schematic of ice maker

automatically selected. A crusher breaks the cubed ice in the crushed mode. Normally, the mechanical load of the icemaker is low because it is operated without fused or webbed ice.

However, for Asian customers, fused or webbed ice will frequently form in the tray because they dispense ice in cubed mode infrequently. When ice is dispensed under these conditions, a serious mechanical overload occurs in the ice crusher. However, in the United States or Europe, the icemaker system operates continuously as it is repetitively used in both cubed and crushed ice modes. This can produce a mechanical/civil overload.

Figure 5.14 overviews the schematic of the icemaker. Figures 5.15 and 5.16 show a schematic diagram of the mechanical/civil load transfer in the ice bucket assembly and its bond graphs. An AC auger motor generates enough torque to crush the ice. Motor power is transferred through the gear system to the ice bucket assembly—that is, to the helix upper dispenser, the blade dispenser and the ice crusher.

The Bond Graph can be represented as a state equation form, that is,

$$df E_2/dt = 1/L_a \times eE_2 \qquad (5.43)$$

$$df M_2/dt = 1/J \times eM_2 \qquad (5.44)$$

The junction from Eq. (5.43)

$$eE_2 = e_a - eE_3 \qquad (5.45a)$$

$$eE_3 = R_a \times fE_3 \qquad (5.45b)$$

The junction from Eq. (5.44)

$$eM_2 = eM_1 - eM_3 \qquad (5.46a)$$

$$eM_1 = (K_a \times i) - T_{\text{Pulse}} \qquad (5.46b)$$

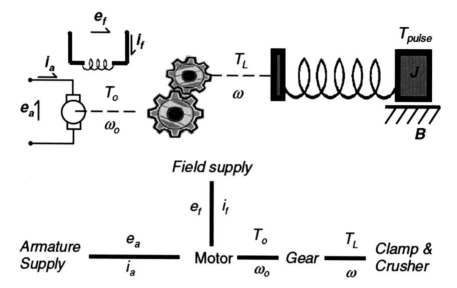

Fig. 5.15 Schematic diagram for mechanical ice bucket assembly

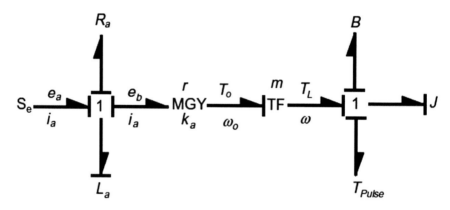

Fig. 5.16 Bond graph of ice bucket assembly

$$eM_3 = B \times fM_3 \qquad (5.46c)$$

Because $fM_1 = fM_2 = fM_3 = \omega$ and $i = fE_1 = fE_2 = fE_3 = i_a$, From Eq. (5.45a)

$$eE_2 = e_a - R_a \times fE_3 \qquad (5.47)$$

$$fE_2 = fE_3 = i_a \qquad (5.48)$$

5.3 Bond Graph Modeling

If substituting Eqs. (5.47) and (5.48) into (5.43), then

$$di_a/dt = 1/L_a \times (e_a - R_a \times i_a) \qquad (5.49)$$

And from Eq. (5.46a) we can obtain

$$eM_2 = [(K_a \times i) - T_{\text{Pulse}}] - B \times fM_3 \qquad (5.50\text{a})$$

$$i = i_a \qquad (5.50\text{b})$$

$$fM_3 = fM_2 = \omega \qquad (5.50\text{c})$$

If substituting Eq. (5.50a) into (5.44), then

$$d\omega/dt = 1/J \times [(K_a \times i) - T_{\text{Pulse}}] - B \times \omega \qquad (5.51)$$

So the state equation can be obtained from Eq. (5.49) and (5.51) as following

$$\begin{bmatrix} di_a/dt \\ d\omega/dt \end{bmatrix} = \begin{bmatrix} -R_a/L_a & 0 \\ mk_a & -B/J \end{bmatrix} \begin{bmatrix} i_a \\ \omega \end{bmatrix} + \begin{bmatrix} 1/L_a \\ 0 \end{bmatrix} e_a + \begin{bmatrix} 1 \\ -1/J \end{bmatrix} T_{\text{Pulse}} \qquad (5.52)$$

When Eq. (5.52) is integrated, the angular velocity of the ice bucket mechanical assembly is obtained as

$$y_p = \begin{bmatrix} 0 & 1 \end{bmatrix} \begin{bmatrix} i_a \\ \omega \end{bmatrix} \qquad (5.53)$$

5.4 Load Spectrum and Rain-Flow Counting

5.4.1 Introduction

As seen in previous sections, we know that product subjected to a variety of loads can be simulated through dynamics modeling like bond graph. On the other hands, to experimentally measure the load over time, strain gage type transducers are attached to the critical areas of the component. The acquired data from the transducers are usually recorded and stored by a computer or by other devices. After the recorded data is filtered to isolate the primary loads from noise, the recorded data converted from the strain values to torque are counted by rain-flow counting methods. After simplifying the fatigue damage computations, we can apply the Miner's rule (see Fig. 5.17).

With the measurement data over time, we can perform a peak and trough detection to find the turning points in the data. This is known as rain-flow counting. The output of this calculation is called the torque count statistics. Some engineers

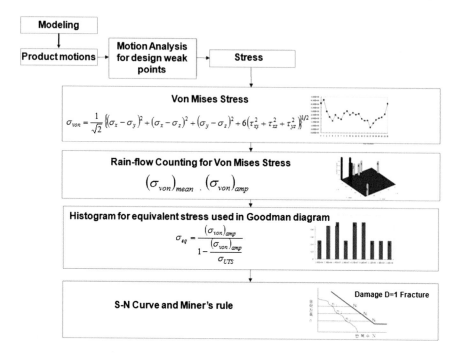

Fig. 5.17 Classification and counting of the dynamic load

stop at this point and define the rain-flow data as the load spectrum, however it is not. Using the rain-flow data, it is then possible to calculate the histogram. This histogram is the load spectrum. This load spectrum is very import during the design phase or a refinement phase. The information from the load spectrum can be used with test rigs or simulation software to reduce, but not remove, the need for field tests.

Realistic representation of loads is a key ingredient to successful fatigue analysis & design. It will accurately measure the applied loads on an existing product and predict loads on a component or structure that does not yet exist. Historically, complex load histories are often replaced by more simplified loadings. The rain-flow cycle counting is a method for counting fatigue cycles from a time history. The fatigue cycles are stress-reversals. The rain-flow method allows the application of Miner's rule in order to assess the fatigue life of a structure subject to complex loading. And rain-flow counting method may enable cumulative damage or the fatigue effects of loading events. The term "spectrum" in fatigue often means a series of fatigue loading events other than uniformly repeated cycles. Sometimes spectrum means a listing, ordered by size, of components of irregular sequences. Maximum and minimum loads are also used to define the classifications in which the counts of cycles are listed.

5.4.2 Rain-Flow Counting

With the load-time, stress-time, or strain-time history, rain flowing down a roof can be represented by the history of peaks and valleys. Rain-flow counting is a concept developed in Japan by Tatsuo Endo and M. Matsuishi in 1968 [1] and in the USA for the segmentation of any arbitrary stress curve into complete oscillation cycles. Rain-flow counting counts closed hysteresis loops in a load-time-function, which are decisive for the damage of metal materials.

The following assumptions are valid for rain-flow counting

- Cyclic stable material behavior, that means that the cyclic stress–strain curve remains constant, thus no hardening or softening of the material takes place.
- Validity of the massing hypothesis, which means that the form of the hysteresis loop branches correspond to the double of the initial load curve.
- Memory behavior of the material which means that after a closed hysteresis loop, a previously not yet completely closed hysteresis loop follows the same σ, ε path.

As seen in Fig. 5.18, the tips of the largest hysteresis loop are at the largest tensile and compressive loads in the load history (points 1 and 4). The notch strain-time history (Fig. 5.18c) is quite different from the corresponding notch stress-time history (Fig. 5.18e). During each segment of the loading the material "remembers" its prior deformation (called material memory). The damage from each counted cycle can be computed from the strain amplitude and mean stress for that cycle as soon as it has been identified in the counting procedure. The corresponding reversal points can then be discarded.

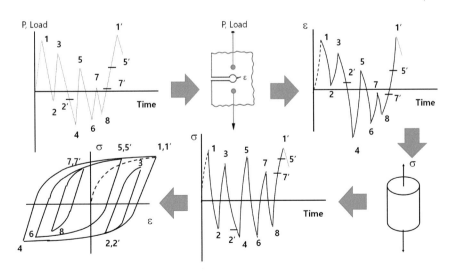

Fig. 5.18 Rain-flow counting method

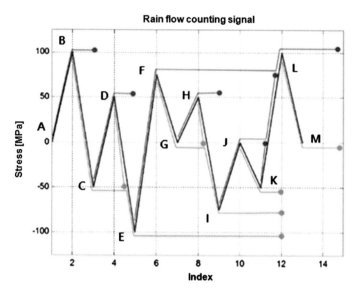

Fig. 5.19 Acquisition of the dynamic load-time behavior with rain-flow counting algorithm

Table 5.3 Summary of the dynamic loads by using rain-flow counting

Path	From (MPa)	To (MPa)	Range (MPa)	Cycles
A–B	0	100	100	0.5
B–E	100	−100	200	0.5
C–D	−50	50	100	0.5
D–C	50	−50	100	0.5
E–F	−100	100	200	0.5
F–I	75	−75	150	0.5
G–H	0	50	50	0.5
H–G	50	0	50	0.5
K–J	−50	0	50	0.5
J–K	0	50	50	0.5
I–F	−75	75	150	0.5
L–M	100	0	100	0.5

That sequence clearly has 10 cycles of amplitude 10 MPa and a structure's life can be estimated from a simple application of the relevant S–N curve.

An advantage of rain-flow counting is when it is used with notch strain analysis. The rain-flow counting results in closed hysteresis loops, which representing a counted cycle. Therefore, the closed hysteresis loops can also be used to obtain the cycle counting. If the dynamic load-time behaviors are acquired in Fig. 5.19, they can be summarized by rain-flow counting as Table 5.3.

- Half-cycle starts at (A) and terminates opposite a greater tensile stress, peak (B); its range is 100 MPa.
- Half-cycle starts at tensile peak (B), flow through (C), and terminates a greater tensile stress, peak (E); its range is 200 MPa.

Consequently, as seen in Table 5.3, we can count two cycles for 50 MPa range, two cycles for 100 MPa range, one cycle for 150 MPa range, and one cycle for 200 MPa range. Since calculated lifetime estimations are afflicted with large uncertainties, it is desired to reconstruct the stochastic load-time functions out of the load spectrums, in order to carry out experimental lifetime proofs with servo-hydraulic facilities.

However, the reconstruction of a representative load-time function is not possible with the load spectra alone. Two parametric rain-flow counting method is the most suitable method for the acquisition of the local stress-strain hysteresis curves and influences the result of lifetime estimation.

5.4.3 Goodman Relation

In the presence of a steady stress superimposed on the cyclic loading, the Goodman relation [2] can be used to estimate a failure condition. It plots stress amplitude against mean stress with the fatigue limit and the ultimate tensile strength of the material as the two extremes.

$$\sigma_a = \sigma'_e \times \left(1 - \frac{\sigma_m}{\sigma'_u}\right), \tag{5.54}$$

where σ'_e effective alternating stress at failure for a lifetime of N_f cycles, σ'_u is ultimate stress.

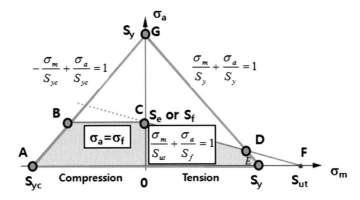

Fig. 5.20 "Augmented" modified-Goodman diagram

A very substantial amount of testing is required to obtain as S–N curve for the simple case of fully reversed loading, and it will usually be impractical to determine whole families of curves for every combination of mean and alternating stress. There are a number of stratagems for finessing this difficulty, one common one being the "Augmented" Modified-Goodman diagram (Fig. 5.20).

Here, a graph is constructed with mean stress as the abscissa and alternating stress as the ordinate, and a straight "lifeline" is drawn from σ_e on the σ_a axis to the ultimate tensile stress σ_f on the σ_m axis. Then for any given stress, the endurance limit (or fatigue limit)—the value of alternating stress at which fatigue facture never occurs—can be read directly as the ordinate of the lifeline at line is drawn from the origin with a slope equal to that ratio. Its intersection with the lifeline then gives the effective endurance limit for that combination of σ_f and σ_m.

5.4.4 Palmgren-Miner's Law for Cumulative Damage

Fatigue properties of a material (S–N curves) are tested in rotating-bending tests in fatigue testing apparatus. The S–N curve is required as a description of the material behavior for the calculation of fatigue strength and operational fatigue strength. Well before a microstructural understanding of fatigue processes was developed, engineers had developed empirical means of quantifying the fatigue process and designing against it. Perhaps the most important concept is the S–N diagram in which a constant cyclic stress amplitude S is applied to a specimen and the number of loading cycles N until the specimen fails. Millions of cycles might be required to cause failure at lower loading levels, so the abscissa in usually plotted logarithmically (Fig. 5.21).

There are three zones to distinguish between in the double logarithmic representation of S–N curve

- Low cycle fatigue: high loads, plastic, and elastic deformation, $N = 10$–10^3 cycles (1 stage)

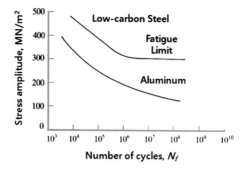

Fig. 5.21 S–N diagram for aluminum and low-carbon steel

5.4 Load Spectrum and Rain-Flow Counting

- High cycle fatigue: fatigue strength, the zone of the sloped lines, until the corner load cycles to failure $N_D = 10^6$–10^7 (2 stage)
- Fatigue limit (endurance limit), zone of the horizontal lines starting from $N > N_D$. However, several materials, such as austenite steels dot possess a distinct endurance strength (3 stage).

In this case, the S–N curve becomes horizontal at large N. The fatigue limit is maximum stress amplitude below which the material never fails, no matter how large the number of cycles is. In most alloys, S decreases continuously with N. In this case, the fatigue properties are described by fatigue strength at which fracture occurs after a specified number of cycles (e.g., 10^7). Fatigue life is number of cycles to fail at a specified stress level.

Fatigue failure has three stages: (1) crack initiation in the areas of stress concentration or near stress raisers, (2) incremental crack propagation, and (3) final rapid crack propagation after crack reaches critical size. The total number of cycles to failure is the sum of cycles at the first and the second stages. That:

$$N_f = N_i + N_p, \tag{5.55}$$

where N_f number of cycles to failure, N_i Number of cycles for crack initiation, N_p Number of cycles for crack propagation

In the fatigue strength zone, the S–N curve can be described by the following equation if represented in the double logarithmic form.

$$N = N_D \cdot \left(\frac{\sigma_a}{\sigma_D}\right)^{-k} \tag{5.56}$$

If possible, the determination of the S–N curve for operational fatigue strength calculation should be carried out on real parts. Often, however, due to cost and time limitations, the calculations are only carried out on special test samples.

The resulting load cycles to failure are random variables, which mean that they lie scattered around the mean value. Today, the transformation of results won from a tension/compression trial onto a real component is difficult. Thus, the exact determination of a notch over the entire load cycle zone is still not possible today. Therefore, one is forced to rely on tests and trials.

In some materials, notably ferrous alloys, the S–N curve flattens out eventually, so that below a certain fatigue limit σ_e failure does not occur no matter how long the loads are cycled. Obviously, the designer will size the structure to keep the stresses below σ_e by a suitable safety factor if cyclic loads are to be withstood. For some other materials, such as aluminum, no fatigue limit exists and the designer will size the structure to keep the stresses below σ_e by a suitable safety factor if cyclic loads are to be withstood. For some other materials such as aluminum, no fatigue limit exists and the designer must arrange for the planned lifetime of the structure to be less than the fatigue point on the S–N diagram.

Statistical variability is troublesome in fatigue testing; it is necessary to measure the lifetimes of perhaps twenty specimens at each of ten or so load levels to define the S–N diagram with statistical confidence. It is generally impossible to cycle the specimen at more than approximately 10 Hz and at that speed it takes 11.6 days to reach 10^7 cycles of loading. Obtaining a full S–N curve is obviously a tedious and expensive procedure.

At first glance, the scatter in measured lifetimes seems enormous, especially given the logarithmic scale of the abscissa. If the coefficient of variability in conventional tensile testing is usually only a few percent, why do the fatigue lifetimes vary over orders of magnitude? It must be remembered that in tensile testing, we are measuring the variability in cycles at a given number of cycles, while in fatigue we are measuring the variability in cycles at a given stress. State differently, in tensile testing we are generating vertical scatters bars, but in fatigue they are horizontal. Note that we must expect more variability in the lifetimes as the S–N curve becomes flatter, so that materials that are less prone to fatigue damage require more specimens to provide a given confidence limit on lifetime.

Numerous different researchers have occupied themselves with the damage accumulation hypothesis in fatigue failure, so that currently several variations exist. In general, the variations only distinguish themselves by the fundamental S–N curve used: either fictitiously extrapolated or the real curve itself.

Oscillating loads cause an effect in materials, this is often referred to as "Damage" as soon as this load surpasses a certain limit. It is assumed that this damage accumulates from the individual load cycles and leads to a material fatigue. For an exact calculation, this damage must be collected and recorded quantitatively. This, however, has not yet been achieved with success.

Despite this fact, in order to gather information concerning the lifetime L out of the results of Wöhler trials with irregular load cycle effects, around the year 1920, Palmgren [3] developed the fundamental idea of linear accumulation, specific for roll bearing calculation. In 1945, *Miner* published the same idea in a general form.

Miner assumes that a part absorbs work during the fatigue process (Fig. 5.22). The ratio of already absorbed work to the maximal work, which can be absorbed is a measurement for the current damage. Thus, the ratio of the load cycle number n to the load cycles to failure N, which is determined in the single-stage zone with the corresponding amplitude, is equal to the ratio of absorbed work w to absorbed work W. This is denoted as the damage portion

$$\frac{w}{W} = \frac{n}{N} \tag{5.57}$$

When the cycle load level varies during the fatigue process, a cumulative damage model is often hypothesized. By definition of the S–N curve, take the lifetime to N_1 cycles at stress level S_1 and N_2 at S_2. If damage is assumed to accumulate at a constant rate during fatigue and a number of cycles n_1 is applied at stress S_1, where $n_1 < N_1$, then the fraction of lifetime consumed will be n_1/N_1.

5.4 Load Spectrum and Rain-Flow Counting

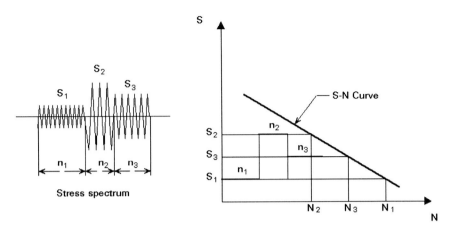

Fig. 5.22 Schematic representation of Miner's cumulative damage summation

The Palmgren-Miner hypothesis asserts that the damage fraction at any level S_i is linearly proportional to the ratio of number of cycles of operation to the total number of cycles that would produce failure at that stress level; that is

$$D_i = \frac{n_j}{N_j} \tag{5.58}$$

The limiting condition of strength happens when the absorbed work and absorbable work are the same. That is, the prerequisite that the absorbed fracture work W is the same for all occurring load sizes, allows the addition of the individual damage portions for load cycles of different sizes

$$\frac{w_1 + w_2 + \cdots + w_m}{W} = 1 \tag{5.59}$$

So failure is predicted as follows,

$$D_1 + D_2 + \cdots + D_i = \frac{w_1}{W} + \frac{w_2}{W} + \cdots + \frac{w_m}{W} = \frac{n_1}{N_1} + \frac{n_2}{N_2} + \cdots + \frac{n_m}{N_m} = 1 \tag{5.60}$$

The generalization of this approach is called Palmgren-Miner's Law, and can be written

$$\sum \frac{n_j}{N_j} \leq 1 \tag{5.61}$$

where n_j is the number of cycles applied at a load corresponding to a lifetime of N_j

Miner confined the applicability of this equation by the following conditions

- Sinus formed load curve
- No hardening or softening appearances in the material
- The begin of a crack is considered as an incipient damage
- Some loads lie above the endurance strength

Minor's law should be viewed like many other material laws that might be accurate enough to use in design. But damage accumulation in fatigue is usually a complicated mixture of several different mechanisms, and the assumption of linear damage accumulation inherent in Miner's law should be viewed skeptically. If portions of the material's microstructure become unable to bear load as fatigue progresses, the stress must be carried by the surviving microstructural elements. The rate of damage accumulation could drop during some part of the material's lifetime. Miner's law ignores such effects, and often fails to capture the essential physics of the fatigue process

With knowledge of the load spectrum and the tolerable material load in the form of the S–N curve, a lifetime prediction can be made for a mechanical/civil system with the help of a damage accumulation hypothesis. Here it should be considered, that this prediction can only be made with a certain probability, since among other things the load spectrum as well as the load capacity expressed in the form of S–N curve are random variables. Likewise, the damage accumulation hypotheses known today have only been proven empirically in material science. Therefore, a practical lifetime prediction requires balance field tests, test stand trials, calculation and a careful assessment, and evaluation of the data, if the prediction should be able to serve as an effective tool for the designer.

Example 5.1 Stress σ_1 has lifetime $N_1 = 10^4$ cycles, and a more rigorous stress σ_2 has lifetime $N_2 = 10^3$ cycles. If 700 cycles at stress σ_2 is operated, when will it stop to operate at stress σ_1?

Solution From Palmgren-Miner's Law Eq. (5.61), we can calculate the cycles to fail.

$$\frac{700}{1000} + \frac{x}{10,000} = 1$$

So the expected failure cycle is $x = 3000$ cycles

Example 5.2 A part is subjected to a fatigue environment where 10% of its life is spent at an alternating stress level, σ_1, 30% is spent at a level σ_2, and 60% at a level σ_3. How many cycles, n, can the part undergo before failure?

If, from the S–N diagram for this material the number of cycles to failure at σ_1 ($i = 1, 2, 3$), then from the Palmgren-Miner rule failure occurs when:

5.4 Load Spectrum and Rain-Flow Counting

$$\frac{0.1n}{N_1} + \frac{0.3n}{N_2} + \frac{0.6n}{N_3} = 1$$

so solving for *n* gives

$$n = \frac{1}{\frac{0.1}{N_1} + \frac{0.3}{N_2} + \frac{0.6}{N_3}}$$

If N_1, N_2, N_3 are $10^3, 10^4$, and 10^5, the time to failure n will be 7353 cycles.

References

1. Matsuishi M, Endo T (1968) Fatigue of metals subjected to varying stress. Japan Soc Mech Eng
2. Mott, R. L. (2004). Machine elements in mechanical design, 4th edn. Pearson Prentice Hall, Upper Saddle River, pp 190–192
3. Palmgren AG (1924) Die Lebensdauer von Kugellagern Zeitschrift des Vereines Deutscher Ingenieure 68(14):339–341

Chapter 6
Mechanical System Failures

Abstract This chapter will review the concepts of fracture and fatigue that occupy most of part failures in mechanical system subjected to random stress (or loads). To figure out the mechanical system failures, it will benefit engineer to design the product structures—automobile, bridge, skyscrapers, and the others—in the allowable stress and strain as mechanical properties. However, as the current reliability methodology, engineer still does not know whether the stress in product lifetime overcomes the random stress in lifetime. Failure of mechanical components in aircraft wing during a long flight can occur in short time or tens of thousands of vibration load cycles. Fatigue fracture catastrophically occurs in product lifetime when there are stress raisers such as holes, notches, or fillets in design. Mechanical engineer should clearly figure out the failure mechanism of fracture or fatigue to design the product structure subjected to random loading. Consequently, to discover the problematic part, we need new reliability methodology like parametric accelerated life testing in the reliability-embedded design process.

Keywords Mechanical system failures · Fracture · Fatigue · Design · Failure analysis

6.1 Introduction

Fatigue was coined by France engineer Jean-Victor Poncelet in the middle of the nineteenth century. It meant that the material got tired due to repeated loading, and eventually disintegrated [1]. The National Bureau of Standards and Battelle Memorial Institute estimated the costs for failure due to fracture to be $119 billion per year in 1982 [2]. The required costs are important, but the safety of many failures in human life and injury is infinitely more.

Fracture mechanics is the study field concerned how cracks in materials propagate. It uses methods of analytical solid mechanics to calculate the driving force on a crack and those of experimental solid mechanics to characterize the material's resistance to fracture. When subjected to a variety of loading, fractures have

Fig. 6.1 Definition of stress

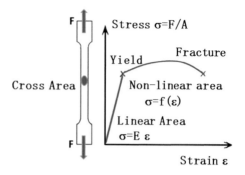

occurred in design inadequacies. Design against fracture still has a area of current research and its final goal.

As seen in Fig. 6.1, stress is a physical quantity that expresses the response of the material on the unit area (*A*) acted in the external (or internal) forces (*F*). And strain is physical deformation response of a material to stress.

$$\sigma = \frac{F}{A} \tag{6.1}$$

The linear portion of the stress–strain curve is the elastic region and the slope is Young's modulus. The elastic range ends when the material reaches its yield strength. After the yield point, the curve typically decreases slightly and deformation continues. Strain hardening and plastic deformation begin when it reaches the ultimate tensile stress.

Deformation refers to any changes in the shape of an object due to an applied force. Elastic deformation is that once the forces are no longer applied, the object returns to its original shape. This type of deformation involves stretching of the atoms bonds. Linear elastic deformation is governed by Hooke's law, which states:

$$\sigma = E\varepsilon \tag{6.2}$$

where σ is the applied stress, E is Young's modulus, and ε is the strain.

Applied force consists of tension, compression, shear, and torsion (Fig. 6.2). Tensile means the material is under tension. The forces acting on it are trying to stretch the material. Compression is when the forces acting on an object are trying to squash it.

- Axial loading (tension/compression)—The applied forces are collinear with the longitudinal axis of the member. The forces cause the member to either stretch or shorten.
- Transverse loading (shear)—Forces are applied perpendicular to the longitudinal axis of a member. Transverse loading causes the member to bend and deflect from its original position, with internal tensile and compressive strains accompanying the change in curvature of the member. Transverse loading also

6.1 Introduction

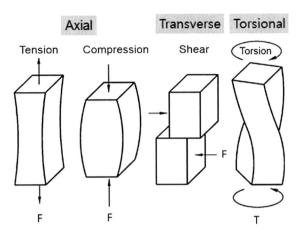

Fig. 6.2 A variety of stresses due to four kinds of static loads

induces shear forces that cause shear deformation of the material and increase the transverse deflection of the member.

- Torsional loading—Twisting action caused by a pair of externally applied equal and oppositely directed force couples acting on parallel planes or by a single external couple applied to a member that has one end fixed against rotation.

Many materials are made up of many grains which may have second phase particles and grain boundaries. It is therefore easier to study plastic deformation in a single crystal to eliminate the effects of grain boundaries and second phase particles.

If a single crystal of a metal is stressed in tension beyond its elastic limit, it elongates slightly, a process known as plastic deformation (Fig. 6.3). Plastic deformation involves the breaking and remaking of atomic bonds. Plastic deformation may take place by slip, twinning, or a combination of both methods. Plastic

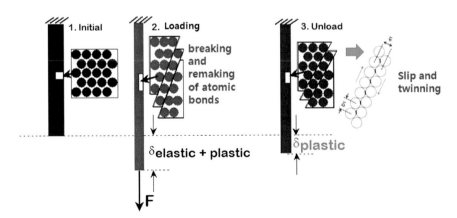

Fig. 6.3 Plastic deformation (metals) for axial force

deformation cannot be restored to its initial state by changes, i.e., irreversible process. Under tensile stress, plastic deformation is characterized by a strain hardening region, necking region, and finally, fracture (also called rupture).

During strain hardening, the material becomes stronger through the movement of atomic dislocations. The necking phase is indicated by a reduction in cross-sectional area of the specimen. Necking begins after the ultimate strength is reached. During necking, the material can no longer withstand the maximum stress, and the strain in the specimen rapidly increases. Plastic deformation ends with the fracture of the material. Fracture is the separation of a single body into pieces by an applied stress.

6.2 Mechanism of Slip

Slip occurs on planes that have highest planer density of atoms and in the direction with highest linear density of atoms (Fig. 6.4). That is, slip occurs in directions in which the atoms are most closely packed since this requires the least amount of energy. Therefore they can slip past each other with less force. Slip flow depends upon the repetitive structure of the crystal which allows the atoms to shear away from their original neighbors. It therefore slides along the face and joins up with the atom of new crystals.

Slip takes place as a result of simple shearing stress. Resolution of axial tensile load F gives two loads. F_s is shear load along the slip plane and the other F_N a normal tensile load perpendicular to the plane. By analysis and experiment, maximum shear stress happens at 45°. Figure 6.4 in the right side shows the packing of

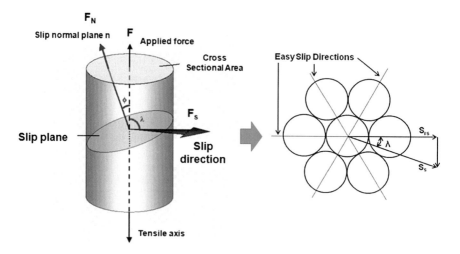

Fig. 6.4 Components of force on a slip plane

6.2 Mechanism of Slip

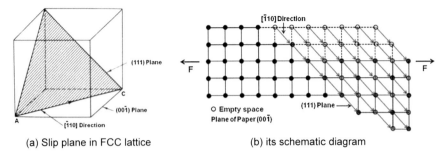

Fig. 6.5 Schematic diagram of slip plane in FCC lattice

atoms on a slip plane. We know that there are three directions in which the atoms are closely packed, and these would be the easy slip directions.

Portions of the crystal on either side of a specific slip plane move in opposite directions and come to rest with the atoms in nearly equilibrium positions, so that there is very little change in the lattice orientation. Thus the external shape of the crystal is changed without destroying it. Schematically, slip can be explained in a face-centered cubic (FCC) lattice. The (111) plane is the slip plane having maximum number of atoms (densest plane). It intersects the (001) plane in the line AC, (110) direction having maximum number of atoms on it. Slip is seen as a movement along the (111) planes in the close-packed (110) direction (Fig. 6.5a).

From the schematic diagram of slip in a FCC crystal, one may assume that the atoms slip consecutively, starting at one place or at a few places in the slip plane, and then move outward over the rest of the plane. For instance, if one tries to slide the entire rug as one piece, the resistance is too much. What one can do is to make a wrinkle in the rug and then slide the whole rug a little at a time by pushing the wrinkle along. A similar analogy to the wrinkle in the rug is the movement of an earthworm. It advances in a direction by advancing a part of its body at a time.

By application of the shear force, first an extra plane of atoms (called a dislocation) forms above the slip plane. On further application of force, bond between atoms breaks and creates a new bond between atoms and a dislocation. On continued application of force, this dislocation advances by breaking old bonds and making new bonds. In the next move, bond between atoms is broken and a new bond is made between atoms, resulting in a dislocation. Thus, this dislocation moves across the slip plane and leaves a step when it comes out at the surface of the crystal. Each time the dislocation moves across the slip plane, the crystal moves the distance of one atom spacing (Fig. 6.5b).

6.3 Facture Failure

Fracture is the separation of a body into pieces subjected to stress. Fracture takes place whenever the applied loads (or stresses) are more than the resisting strength of the body. It starts with a crack that breaks without making fully apart. Fracture due to overstress is probably the most prevalent failure mechanism in mechanical/civil system and might be classified as ductile fracture, brittle fracture, and fatigue fracture.

As seen in Fig. 6.6, brittle fracture is the failure of a material with minimum of plastic deformation. Brittle fracture propagates rapidly on a crack with minimum energy absorption and plastic deformation. Brittle fracture occurs along characteristic crystallographic planes called as cleavage planes. The mechanism of Brittle fracture was initially explained by Griffith theory [3]. Griffith postulated that in a brittle material there are micro-cracks which act to the concentrated stress at their tips. The crack could come from a number of sources as flow occurred during solidification or a surface scratch.

Brittle materials are glasses, ceramics, some polymers, and metals. They have the following characteristics:

- No appreciable plastic deformation
- Crack propagation is very fast
- Crack propagates nearly perpendicular to the direction of the applied stress
- Crack often propagates by cleavage—breaking of atomic bonds along specific crystallographic planes (cleavage planes).

Ductile fracture occurs after considerable plastic deformation. The crack will not extend unless an increased stress is applied. The failure of most polycrystalline ductile materials occurs with a cup-and-cone fracture associated with the formation of a neck in a tensile specimen. In ductile material, the fracture begins by the formation of cavities (micro-voids) in the center of the necked region. In most commercial metals, these internal cavities probably form at nonmetallic inclusions.

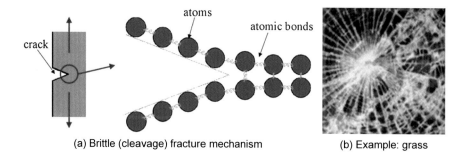

(a) Brittle (cleavage) fracture mechanism   (b) Example: grass

Fig. 6.6 Brittle (cleavage) fracture mechanism

6.3 Facture Failure

Increasing the load, increases the permanent elongation and simultaneously decreases the cross-sectional area. The decrease in area leads to the formation of a neck in the specimen.

The neck region has a high dislocation density and the material is subjected to a complex stress. The dislocations are separated from each other because of the repulsive interatomic forces. As the resolved shear stress on the slip plane increases, the dislocation comes closely together. The crack forms due to high shear stress and in the presence of low angle grain boundaries. Once a crack is formed, it can grow or elongate by means of dislocations. Crack propagates along the slip plane for this mechanism. Once crack grows at the expense of others, finally crack's growth results in failure. The final stage leaves a circular lip on one half of the sample and a bevel on the surface of the other half. Thus one half has the appearance of a shallow cup, and the other half resembles a cone with a flattened top (see Figs. 6.7 and 6.8).

- Brittle Fracture: Separation along crystallographic planes due to breaking of atomic bonds (V-shaped Chevron, Cleavage, Inter-granular)
- Ductile Fracture: Initiation, growth, and coalescence of micro-voids (Cup-and-cone, Dimple).

However, commercial material is made up of polycrystalline, whose crystal axes are oriented at random. When polycrystalline material is subjected to stress, slip starts first in those grains in which the slip system is most favorably situated with respect to the applied stress. Since contact at the grain boundaries is maintained, it may be necessary for more than one slip system to operate. The rotation into the axis of tension brings other grains, originally less favorably oriented, into a position where they can now deform. As deformation and rotation proceed, the individual grains tend to elongate in the direction of flow.

When a crystal deforms, there is some distortion of the lattice structure. This deformation is greatest on the slip planes and grain boundaries and increases with deformation. This is evident by an increase in resistance to further deformation.

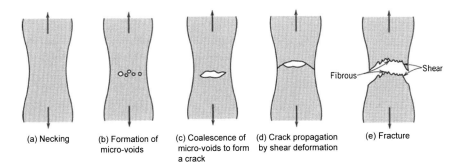

(a) Necking (b) Formation of micro-voids (c) Coalescence of micro-voids to form a crack (d) Crack propagation by shear deformation (e) Fracture

Fig. 6.7 Ductile fracture failure mechanism

Fig. 6.8 Brittle versus ductile fracture in material

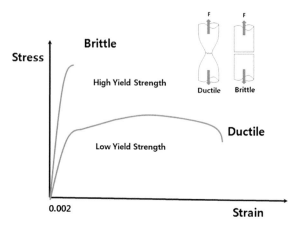

The material is undergoing strain hardening or work hardening. Since dislocations pile up at grain boundaries, metals can be hardened by reducing the size of the grains.

6.4 Fatigue Failure

6.4.1 Introduction

Another deformation mechanism is fatigue failure, which occurs primarily in ductile metals. Fatigue may occur when a member is subjected to repeated cyclic loadings. The fatigue phenomenon shows itself in the form of cracks developing at particular locations in the structure. Cracks can appear in diverse types of structures such as: planes, boats, bridges, frames, cranes, overhead cranes, machines parts, turbines, reactors vessels, canal lock doors, offshore platforms, transmission towers, pylons, masts, and chimneys.

Design faults such as stress raisers are deformed. After repetitive deformations, cracks will begin to appear. Depending on the material, shape, and how close to the elastic limit it deforms, failure may require a lot of deformation cycles. Structures subjected to repeated cyclic loadings can undergo progressive damage, which is called fatigue (Fig. 6.9).

The fatigue life of a structural detail subjected to repeated cyclic loadings is defined as the number of stress cycles it can stand before failure. The physical effect of a repeated load on a material is different from the static load. Failure always may be brittle fracture regardless of whether the material is brittle or ductile. Mostly fatigue failure occurs at stress well below the static elastic strength of the material.

Depending upon the member or structural detail geometry, its fabrication or the material used, four main parameters can influence the fatigue strength: (1) the stress difference, or as most often called stress range, (2) the structural geometry, (3) the material, (4) the environment

6.4 Fatigue Failure

Fig. 6.9 Facture of train wreck due to metal fatigue failure of rail from Wikipedia

6.4.2 Type of Fatigue Loading

Cyclic loading due to repeated force and weight of product is a universal loading condition. Essentially all structural components are subjected to some type of fluctuating loading during product lifetime, so they develop fatigue inducing varying stresses. It is virtually impossible to think of any structural component that does not experience some form of alternating loading.

Three different fluctuating stress–time modes are symmetrical about zero stress, asymmetrical about zero stress, and random stress cycle. For reversed stress cycle, amplitude is symmetric about a mean zero stress level. It alternates from σ_{max} to σ_{min} of equal magnitude. Repeated stress cycle is asymmetrical about σ_{max} and σ_{min} relative to zero stress level. Random stress cycle is that stress level fluctuates very randomly in amplitude and frequency.

For asymmetrical about zero stress, cyclic stresses that arise fatigue are characterized by mean stress σ_m, the range of stress $\Delta\sigma$, alternating component σ_a, amplitude ration A, and the stress ratio R (see Fig. 6.10), They are represented as follows Eqs. (6.3)–(6.7).

$$\sigma_m = \frac{(\sigma_{max} + \sigma_{min})}{2} \quad (6.3)$$

$$\Delta\sigma = (\sigma_{max} - \sigma_{min}) \quad (6.4)$$

Fig. 6.10 Fatigue: Failure under fluctuating and cyclic stresses asymmetrical about zero stress

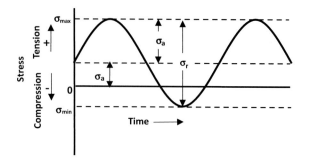

$$\sigma_a = \frac{(\sigma_{max} - \sigma_{min})}{2} \qquad (6.5)$$

$$A = \frac{\sigma_a}{\sigma_m} \qquad (6.6)$$

$$R = \sigma_{min}/\sigma_{max} \qquad (6.7)$$

As seen in Fig. 6.11, other cyclic stresses that arise fatigue are (1) periodic and symmetrical about zero stress, (2) random stress fluctuations. In mechanical/civil system such as bridges, aircraft, machine components, and automobiles, fatigue failure under fluctuating/cyclic stresses are required:

Fig. 6.11 Fatigue: failure under fluctuating and cyclic stresses

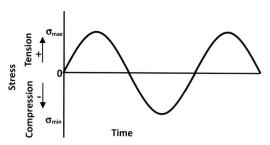

(a) Periodic and symmetrical about zero stress

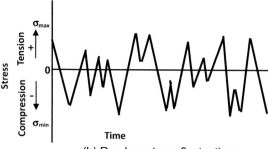

(b) Random stress fluctuations

6.4 Fatigue Failure

- A maximum tensile stress of sufficiently high value
- A large enough variation or fluctuation in the applied stress, and
- A sufficiently large number of applied stress cycles.

There are two ways to determine when the component is in danger of metal fatigue: either predict when failure will occur due to the material/force/shape/iteration combination, and replace the vulnerable materials before this occurs, or perform inspections to detect the microscopic cracks and perform replacement once they occur. Selection of materials not likely to suffer from metal fatigue during the life of the product is the best solution, but not always possible. Avoiding shapes with sharp corners limits metal fatigue by reducing stress concentrations, but does not eliminate it.

Fatigue is a critical failure mechanism to be considered in product designs. It is a process in which damage accumulates due to the repetitive loads below the yield point, which is brittle-like even in normally ductile materials. Fatigue cracks begin very small and initially grow very slowly until the crack length approaches the critical length. So it is dangerous because it is difficult to initially detect cumulative fatigue damage with the naked eye until the crack has grown to near critical length. Typical fracture surface is perpendicular to direction of applied stress. Fatigue failure has three distinct stages: (1) crack initiation in the areas of stress concentration (near stress raisers), (2) incremental crack propagation, and (3) final catastrophic failure.

The examples of "fatigue" for a multitude of reasons have been studied as the disaster of Comet aircraft and Versailles rail accident occurred when they became large enough to propagate catastrophically (see Chap. 2). Fatigue failure occurs in both metallic and nonmetallic materials, and is responsible for about estimated 80–90% of all structural failures—automobile crankshaft, motor shaft, bridges, aircraft landing gear machine components, and the others. Thus, designing for maximum stress will not ensure adequate product lifetime. Most fracture induced belongs to this category.

Engineering stress is irregular around stress raisers such as holes, notches, or fillets that concentrate on the stress. For complex drawings, engineer frequently neglects these design flaws that might cause the reliability disasters. For instance the vibration of aircraft wing during a long flight can result in tens of thousands of load cycles. If designed improperly, these structures will fracture. It is important to find the design faults. In Chap. 7 we will discuss how to find the missing design parameters by using the parametric ALT.

The central difficulty in designing against fracture in high-strength materials is that the presence of cracks can modify the local stresses to such an extent that the elastic stress analyses done so carefully by the designers are insufficient. When a crack reaches a certain critical length, it can propagate catastrophically through the structure, even though the gross stress is much lesser than what would normally cause yield or failure in a tensile specimen. The term "fracture mechanics" refers to a vital specialization within solid mechanics in which the presence of a crack is assumed, and we wish to find quantitative relations between the crack length, the material's inherent resistance to crack growth, and the stress at which the crack propagates at high speed to cause structural failure.

Fast fracture can occur within a few loading cycles. For example, fatigue failures in 1200 rpm motor shafts took less than 12 h from installation to final fracture, about 830,000 cycles. On the other hand, crack growth in slowly rotating process equipment shafts has taken many months and more than 10,000,000 cycles to fail.

6.4.3 Stress Concentration at Crack Tip

Fracture strength of a material is related to the cohesive forces between atoms. One can estimate that the theoretical cohesive strength of a material should be one-tenth of the elastic modulus (E). However, the experimental fracture strength for brittle material is normally $E/100$–$E/10,000$, below this theoretical value. This much lower fracture strength is caused from the stress concentration due to the presence of microscopic flaws or cracks found either on the surface or within the material. As seen in Fig. 6.12, stress profile along X axis is concentrated at an internal, elliptically shaped crack.

Stress has a maximum at the crack tip and decreases to the nominal applied stress with the increasing of distance from the crack. Flaws such stress concentrators or stress raisers have the ability to amplify the stress at a given point. The magnitude of amplification depends on crack geometry and orientation.

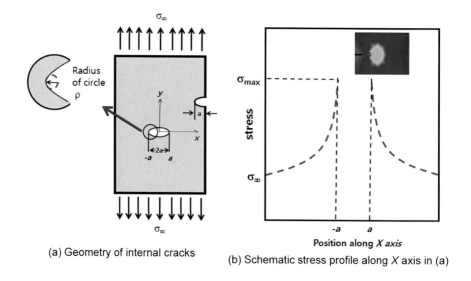

(a) Geometry of internal cracks

(b) Schematic stress profile along X axis in (a)

Fig. 6.12 Stress concentration at crack tip positions

6.4 Fatigue Failure

If the crack is similar to an elliptical hole through plate and is oriented perpendicular to applied stress, the maximum stress σ_{max} occurs at a crack tip and approximated by Eq. (6.8)

$$\sigma_{max} = \sigma_\infty \left[1 + 2\sqrt{\frac{a}{\rho}}\right], \tag{6.8}$$

where ρ = radius of curvature, σ_∞ = applied stress, σ_{max} = stress at crack tip, a = half-length of internal crack or the full length for a surface flaw.

The magnitude of the nominal applied tensile stress is σ_∞; the radius of the curvature of the crack tip is ρ; and a represents the length of a surface crack, or half the length of an internal crack. For a relatively long micro-crack, the factor $(a/\rho)^{1/2}$ may be very large. So Eq. (6.8) can be modified as:

$$\sigma_m \cong 2\sigma_\infty \left(\frac{a}{\rho}\right)^{1/2} \tag{6.9}$$

The ratio of the maximum stress and the nominal applied tensile stress is denoted as the stress concentration factor K_t. The stress concentration factor is a simple measure of the degree to which an external stress is amplified at the tip of a small crack and described as:

$$K_t = \frac{\sigma_{max}}{\sigma_o} \approx 2\left(\frac{a}{\rho_t}\right)^{1/2} \tag{6.10}$$

Because an external stress is amplified at the tip of a crack, Eq. (6.10) can be rearranged as:

$$\sigma_{max} = 2\sigma_\infty \left(\frac{a}{\rho}\right)^{1/2} = K_t \sigma_\infty \tag{6.11}$$

Stress amplification not only occurs at small flaws or cracks on a microscopic level of material but can also occur in sharp corners, holes, fillets, and notches on the macroscopic level. Cracks with sharp tips propagate easier than cracks having blunt tips. Because of amplifying an applied stress, stress concentration may occur at microscopic defects, internal discontinuities (voids/inclusions), sharp corners, scratches, and notches that are often called stress raisers. Stress raisers are typically more destructive in brittle materials. Ductile materials have the ability to plastically deform in the region surrounding the stress raisers which in turn evenly distribute the stress load around the flaw. The maximum stress concentration factor results in a value less than that found for the theoretical value. Since brittle materials cannot plastically deform, the stress raisers will create the theoretical stress concentration situation.

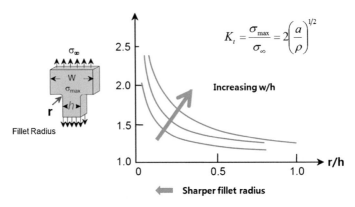

Fig. 6.13 Stress concentration at sharp corners in accordance with fillet radius [4]

The magnitude of this amplification depends on micro-crack orientations, geometry, and dimensions. For example, stress concentration at sharp corners depends on fillet radius (see Fig. 6.13).

6.4.4 Crack Propagation and Fracture Toughness

Cracks with sharp tips propagate easier than cracks having blunt tips. In ductile materials, plastic deformation at a crack tip "blunts" the crack. Elastic strain energy is stored in material as it is elastically deformed. This energy is released when the crack propagates and creation of new surfaces requires energy. Critical stress that is required for crack propagation is described as:

$$\sigma_c = \left(\frac{2E\gamma_s}{\pi a}\right)^{1/2}, \tag{6.12}$$

where γ_s = specific surface energy

When the tensile stress at the tip of crack exceeds the critical stress value, the crack propagates and results in fracture. Most metals and polymers have plastic deformation. For ductile materials, specific surface energy γ_s should be replaced with $\gamma_s + \gamma_p$ where γ_p is plastic deformation energy. So Eq. (6.12) can be described as:

$$\sigma_c = \left(\frac{2E(\gamma_s + \gamma_p)}{\pi a}\right)^{1/2} \tag{6.13}$$

6.4 Fatigue Failure

For highly ductile materials, $\gamma_p \gg \gamma_s$ is valid. So Eq. (6.13) can be modified as:

$$\sigma_c = \left(\frac{2E\gamma_p}{\pi a}\right)^{1/2} \quad (6.14)$$

All brittle materials contain a population of small flaws that have variety of sizes. When the magnitude of the tensile stress at the tip of crack exceeds the critical stress value, the crack propagates and results in fracture. Very small and virtually defect-free metallic and ceramic materials have been produced with facture strength that approaches their theoretical values.

Example 6.1 There is a long plate of glass subjected to a tensile stress of 30 MPa. If the modulus of elasticity and specific surface energy for this glass are 70 GPa and 0.4 J/m², find out the critical length of a surface flaw that can have no fracture.

From Eq. (6.12), $E = 70$ GPa, $\gamma_s = 0.4$ J/m², $\sigma = 40$ MPa. So the critical length can be obtained as

$$a_c = \left(\frac{2E\gamma_s}{\pi\sigma^2}\right) = \left(\frac{2.70 \text{ GPa} \cdot 0.4 \text{ J/m}^2}{\pi \cdot (30 \text{ MPa})^2}\right) = 2.0 \times 10^{-6} \text{ m}$$

Fracture toughness K_c is a material's resistance to fracture when a crack is present. It therefore means the amount of stress required to propagate a flaw. It can be described as:

$$K_c = \sigma_c \sqrt{\pi a} \quad (6.15)$$

Fracture toughness depends on temperature, strain rate, and microstructure. Its magnitude diminishes with increasing strain rate and decreasing temperature. If yield strength due to alloying and strain hardening improves, fracture toughness will increase with reduction in grain size.

6.4.5 Crack Growth Rates

The metal fatigue begins at an internal (or surface) flaw by the concentrated stresses, and progress initially of shear flow along slip planes. As previously mentioned in Sect. 6.2, slip can happen (111) plane in a FCC lattice because the atoms are most closely packed. Over a number of random loading cycles in field, this slip generates intrusions and extrusions that begin to resemble a crack. A true crack running inward from an intrusion region may propagate initially along one of the original slip planes, but eventually turns to propagate transversely to the principal normal stress.

After repeated loadings, the slip bands can grow into tiny shear-driven micro-cracks. These Stage I cracks can be described as a back and forth slip on a series of contiguous crystallographic plane to form a band. It is within this slip bands that the process of pores nucleation and coalescence occurs. The process eventually leads to micro-cracks formation. Often, extrusion and intrusions may also appear which, being a very localized discontinuity, results in a much faster micro crack formation.

Micro-cracks join to form a macro-crack in Stage II of fatigue. Now the crack is already long enough to escape shearing stress control and be driven by normal stress which produces a continuous growth, cycle by cycle, on a plane that is no longer crystallographic, but rather normal to external loads. Ahead of this macro-crack, two plastic lobes are generated by stress concentration. The cracks grow perpendicular to the dominant stress and increase dramatically by plastic stresses at the crack tip as seen in Fig. 6.14.

It is vital that engineers be able to predict the rate of crack growth during load cycling in aircraft as well as in other engineering structures, so that the problematic parts be replaced or repaired before the crack reaches a critical length. A great deal of experimental evidence supports the view that the crack growth rate can be corrected with the cycle variation in the stress intensity factor [5]:

$$\frac{da}{dN} = A \, \Delta K^m, \qquad (6.16)$$

where da/dN is the fatigue crack rate per cycle, $\Delta K = K_{min} - K_{max}$ is the stress intensity factor range during the cycle, and A and m are parameters that depend the material, environment, frequency, temperature, and stress ratio.

Fatigue crack propagation rate during Stage II depends on stress level, crack size, and materials. This is sometimes known as the "Paris Law," and leads to plots similar to that shown in Fig. 6.15.

Some specific values of the constants m and A for various alloys are given in Table 6.1. The exponent m is often near 4 for metallic systems, which might be rationalized as the damage accumulation being related to the volume V_p of the plastic zone: since the volume V_p of the zone scales with r_p^2 and $r_p \propto K_I^2$, then $da/dn \propto \Delta K^4$.

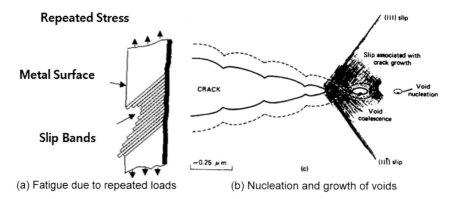

Fig. 6.14 A schematic diagram of general slip producing nucleation and growth of voids

6.4 Fatigue Failure

Fig. 6.15 Paris law for fatigue crack growth rates

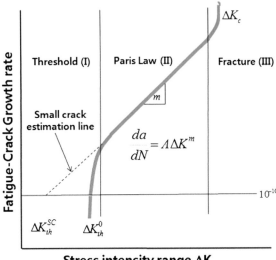

Table 6.1 Numerical parameters in the Paris equation

Alloy	m	A
Steel	3	10^{-11}
Aluminum	3	10^{-12}
Nickel	3.3	4×10^{-12}
Titanium	5	10^{-11}

6.4.6 Ductile–Brittle Transition Temperature (DBTT)

The Ductile-to-Brittle Transition Temperature (DBTT) is widely observed in metals that are dependent on the composition of the metal. For some steels, the transition temperature can be around 0 °C, and in winter the temperature in some parts of the world can be below this. As a result, some steel structures are very likely to fail in winter. The controlling mechanism of this transition still remains unclear despite the large efforts made in experimental and theoretical investigation. All ferrous materials (except the austenitic grades) exhibit a transition from ductile to brittle when tested above and below a certain temperature, called as DBTT. FCC metals such as Cu, Ni remain ductile down to very low temperatures. For ceramics, this type of transition occurs at much higher temperatures than for metals (Fig. 6.16).

Since the famous weld fractures in some US army ships (Liberty Ships, tankers) during World War II are investigated, the ductile-to-brittle transition can be measured by impact testing such as Charpy V-notch testing (Fig. 6.17). The impact energy needed for fracture drops suddenly over a relatively narrow temperature range—temperature of the ductile-to-brittle transition. Primary function of Charpy V-notch testing is to determine whether a material experiences a ductile-to-brittle

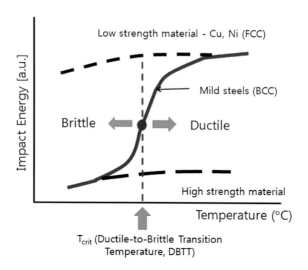

Fig. 6.16 Ductile-to-Brittle transition temperature

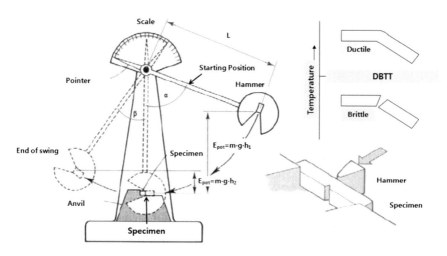

Fig. 6.17 Schematic of a conventional Charpy V-notch testing

transition with decreasing temperature. When the results of a number of tests performed in different temperatures are plotted, ductile-to-brittle transition curves may be obtained.

Steels were used having DBTT's just below room temperature. Low temperatures can severely become brittle steels. At higher temperature, the impact energy is large, corresponding to a ductile mode of fracture. As the temperature is lowered, the impact energy drops suddenly over a relatively narrow temperature range which corresponds to the mode of brittle fracture. Fatigue cracks nucleated at the corners of square hatches and propagated rapidly by brittle fracture.

6.4.7 Fatigue Analysis

The majority of component designs involve parts subjected to fluctuating or cyclic loads. Such loading induces fluctuating or cyclic stresses that often result in failure by fatigue. About 95% of all structural failures occur through a fatigue mechanism. The concept of fatigue that describes structural system subjected to repeated loadings was originated in the mid-eighteenth century in the railroad industry. When fatigue failures of railway axles became a widespread problem in the middle of the nineteenth century, this drew attention to cyclic loading effects. This was the first time that many similar components had been subjected to millions of cycles at stress levels well below the monotonic tensile yield stress.

The modern study of fatigue is generally dated from the work of A. Wöhler, a German engineer in the railroad system in the mid-nineteenth century. Wöhler was chief superintendent of rolling stock on the Lower Silesia–Brandenburg Railroad. Wöhler was concerned by the causes of fracture in rail car axles after prolonged use. A railcar axle is essentially a round beam in four-point bending, which produces a compressive stress along the top surface and a tensile stress along the bottom. After the axle has rotated a half turn, the bottom becomes the top and vice versa, so the stresses on a particular region of material at the surface vary repeatedly form tension to compression. Though the metal became tired, fatigue was named to describe this type of damage. This is now known as fully reversed fatigue loading. At the same time, other engineers began to concern themselves with problems of failures associated with fluctuating loads in bridges, marine equipment, and power generation machines (see Fig. 6.18).

Since 1830, it has been recognized that metal under a repetitive or fluctuating load will fail at a stress level lower than those required to cause failure under a single application of the same load. Figure 6.19 shows a bar-shaped component subjected to a uniform sinusoidally varying force. After a period of time, a crack can be seen to initiate on the circumference of the hole. This crack will then propagate through the component until the remaining intact section is incapable of sustaining the imposed stresses and the component fails.

The physical development of a crack is generally divided into two separate stages. These relate to the crack initiation phase (Stage I) and the crack growth phase (Stage II). Fatigue cracks initiate through the release of shear strain energy. The following diagram shows how the shear stresses result in local plastic deformation along slip planes. As the sinusoidal loading is cycled, the slip planes move back and forth like a pack of cards, resulting in small extrusions and intrusions on the crystal surface. These surface disturbances are approximately 1–10 μm in height and constitute embryonic cracks.

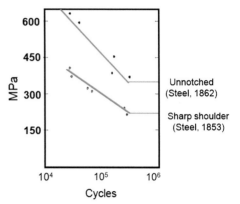

Fig. 6.18 Some of Wöhler's data for rail car axles steel on the S-N diagram [6]

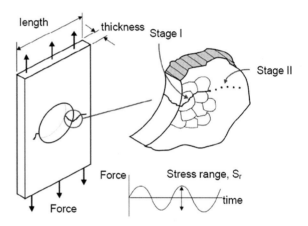

Fig. 6.19 Bar-shaped component subjected to a uniform sinusoidally varying force

A crack initiates in this way until it reaches the grain boundary. The mechanism at this point is gradually transferred to the adjacent grain. When the crack has grown through approximately 3 grains, it is seen to change its direction of propagation. Stage I growth follows the direction of the maximum shear plane, or 45° to the direction of loading. The physical mechanism for fatigue changes during Stage II. The crack is now sufficiently large to form a geometrical stress concentration. A tensile plastic zone is created at the crack tip as shown in the following diagram. After this stage, the crack propagates perpendicular to the direction of the applied load.

6.5 Stress–Strength Analysis

Stress–strength analysis in reliability engineering is the analysis of the strength of the materials and the interference of the stresses placed on the materials [7]. A product's probability of failure is equal to the probability that the stress experienced by that product will exceed its strength. If one probability distribution function for a product's stress and its strength is given, the probability of failure can be estimated by calculating the area of the overlap between the two distributions. This overlapping region may be also referred to as stress–strength interference. However, if there is the design failure like stress raiser in structure, stress–strength interference analysis is not a good expression that can express the root cause of reliability disasters (Fig. 6.20).

If the distributions for both the stress and the strength both follow a normal distribution, the expected probability of failure, F, can be calculated as:

$$F = P[\text{stress} \geq \text{strength}] = \int_0^\infty f_{\text{strength}}(x) \cdot R_{\text{stress}}(x) dx \qquad (6.17)$$

The expected reliability, R, is calculated as:

$$R = P[\text{stress} \leq \text{strength}] = \int_0^\infty f_{\text{stress}}(x) \cdot R_{\text{strength}}(x) dx \qquad (6.18)$$

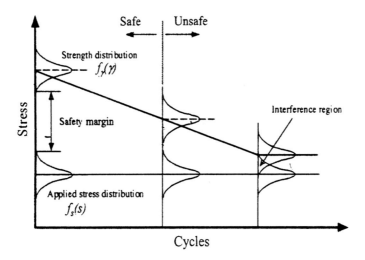

Fig. 6.20 Applied fatigue stress–strength interference model

There are two ways to increase reliability: (1) increase the difference (or safety margin) between the mean stress and strength values, (2) decrease the standard deviations of the distributions of stress and strength. The estimates of stresses and strengths for all component of a product would be perfectly accurate, but this is too costly to accomplish. And the stress conditions depend on the way the product is used—the customer profiles and environmental conditions.

The strength distribution mainly depends on the material used in the product, its dimensions, and the manufacturing process. To improve the product reliability, the product in the design phase should increase its strength by using the optimal design and reliability testing. One method of reliability will be discussed with the parametric ALT testing in Chap. 8.

Environmental stresses have a distribution with a mean μ_x and a standard deviation S_x and component strengths have a distribution with a mean μ_y and a standard deviation S_y. The overlap of these distributions is the probability of failure Z. This overlap is also referred to stress–strength interference.

If stress and strength are normally distributed random variables and are independent of each other, the standard normal distribution and Z tables can be used to quantitatively determine the probability of failure. First, the Z-statistic is calculated as follows:

$$Z = -\frac{\mu_x - \mu_y}{\sqrt{S_x^2 + S_y^2}} \quad (6.19)$$

Using the Z-value table for a standard normal distribution, the area above the calculated Z-statistic is the probability of failure. P(Z) can be determined from a Z table or a statistical software package. For example, if μ_x is 2500 kPa, μ_y is 4500 kPa, S_x is 500 kPa, and S_y is 400 kPa, the probability of failure can be calculated:

$$Z = -\frac{\mu_x - \mu_y}{\sqrt{S_x^2 + S_y^2}} = -\frac{2500 - 4500}{\sqrt{500^2 + 400^2}} = 2.34 \quad (6.20)$$

Using the Z-value table for a standard normal distribution, the area above a Z value of 2.34 (2.34 standard deviations) is 0.0096. Therefore, the probability of failure is 0.96%. Likewise, the reliability is $1 - 0.0096 = 0.9904$ or 99.04%.

6.6 Failure Analysis

6.6.1 Introduction

Using microscopy and spectroscopy, failure analysis is to search out the root cause of failed components in field and to improve product reliability. Failure analysis is

6.6 Failure Analysis

designed to identify the failure modes, the failure site, and the failure mechanism. It determines the root causes of the design and recommends failure prevention methods.

The process begins with the most nondestructive techniques and then proceeds to the more destructive techniques, allowing the gathering of unique data from each technique throughout the process. The sequence of procedures is visual Inspection, electrical testing, nondestructive evaluation, and destructive evaluation.

As seen in Fig. 6.21, failure mechanism of product might be classified as overstress mechanisms and wear mechanisms. Some modes of failure mechanisms are excessive deflection, buckling, ductile fracture, brittle fracture, impact, creep, relaxation, thermal shock, wear, corrosion, stress corrosion cracking, and various types of fatigue. Over time, as more is understood about a failure, the root cause evolves from a description of symptoms and outcomes. The more complex the product or situation, the more necessary a good understanding of its failure cause is to ensuring its proper operation (or repair).

Materials may be degraded by their environment by corrosion processes, such as rusting in the case of iron and steel. Such processes can also be affected by load in the mechanisms of stress corrosion cracking and environmental stress cracking.

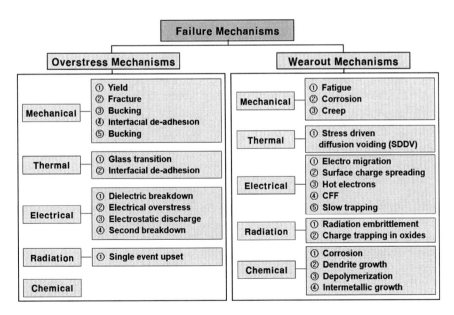

Fig. 6.21 Typical failure mechanisms in product

6.6.2 Procedure of Failure Analysis

To improve reliability targeting of product or modules, the design of a part structure often requires the engineer to minimize the possibility of failure. It therefore is a critical process to understand the failure mechanics—fracture and fatigue. Reliability engineer is familiar with appropriate design principles that can be employed to prevent the failures. By design feedback, reliability engineer can modify the design by correcting the missing design parameters. Manufacturers also need to know "why things fail" as much as they know "how things work."

Failure analysis is a systematic examination of failed products to determine the root cause of failure and to use such information to eventually improve product reliability (see Fig. 6.22). Failure analysis is designed to (1) identify the failure modes (the way the product failed), (2) identify the failure site (where in the product failure occurred), (3) identify the failure mechanism (the physical phenomena involved in the failure), (4) determine the root cause (the design, defect, or loads which led to failure), and (5) recommend failure prevention methods.

It will inspect whether the load applied cyclically or was overload, the direction of the critical load, and the influence of outside forces such as residual stresses. Then, accurately knowing the physical roots of the failure leads to pursue the human errors or the latent causes of these physical roots. Failure analysis might be classified as nondestructive analysis and destructive analysis.

The process begins with the most nondestructive techniques and then proceeds to the more destructive techniques, allowing the gathering of unique data from each technique throughout the process. When properly analyzed, this data leads to a viable failure mechanism. The use of destructive techniques in the early process leads to lose the valuable information that might be required later. The sequence of procedures is:

- Visual Inspection
- Mechanical or Electrical Testing
- Nondestructive Evaluation
- Destructive Evaluation (using relevant techniques).

To increase the product reliability, the results of failure mechanism must be modeled by the physics of failure (PoF). It allows designers to properly select materials, which minimize the susceptibility of future designs to failure by degrading it. In addition, it allows the user to select environmental and operational loads that minimize the susceptibility of the current design to failure during product lifetime.

The identification of the critical failure mechanisms and failure sites of assemblies in field also permits the development of a focused accelerated test program. The benefits of accelerated testing are that it allows the proper test stresses (e.g., temperature, relative humidity, temperature cycling) so as to cause wear-out failure in the shortest time without changing the failure mechanism or mode.

6.6 Failure Analysis

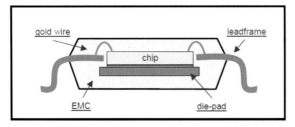

Failure Classification

Physical Failure (Structure)
- Popcorn
- Delamination
- Crack (Package/Die)

Electrical Failure (Connection)
- Open
- Short
- Leakage
- Function

In-Process Failure (Production)
- Front-end (before molding)
- Back-end (After molding)
- Testing (FT/Burn-in)

Reliability Failure (Qualification)
- Temperature
- Humidity
- Pressure
- Voltage

(a) Structure of conventional package in semi-conduct

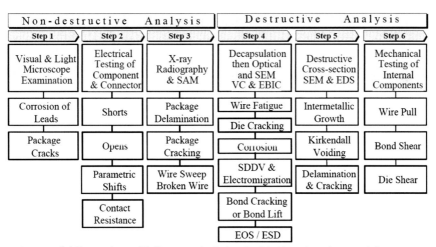

(b) Procedure of failure analysis in electrical system (example)

Fig. 6.22 Failure analysis in electrical system

This is a vast improvement over the old method of choosing a random set of test loads and levels, or subjecting the assemblies to a set of "one size fits all" standard tests prescribed by decades-old military and commercial standards. In addition, the failure distribution in the accelerated tests can be converted to a failure distribution in the intended use environment using the acceleration factors calculated by the PoF models. Typical equipments of failure analysis in product might be used as optical microscope, X-ray, SEM, SAM, FTIR and the others (See Fig. 6.23).

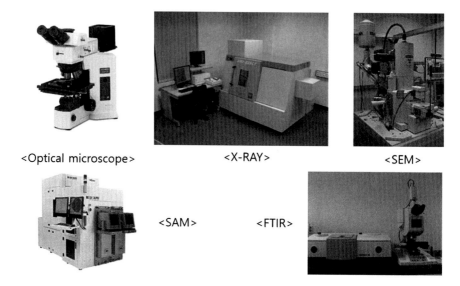

Fig. 6.23 Typical equipments of failure analysis

Nondestructive evaluation (NDE) is designed to provide as much information on the failure site, failure mechanism, and root cause of failure without causing any damage to the product or removing valuable information. A significant amount of failure information is available through visual inspection and the more traditional NDE methods—X-ray or SAM.

For mechanical or electronic device, X-ray microscopy assesses the internal damage, defects, and degradation in microelectronic devices. Illuminating a sample with X-ray energy provides images based on material density that allow characterization of solder voiding, wire-bond sweep, and wire-bond breakage in components. Consequently, X-ray microscopy is a powerful nondestructive tool for pinpointing failure sites in product (see Fig. 6.24).

As destructive evaluation, scanning electron microscopy (SEM) is a natural extension of optical microscopy. The use of electrons instead of a light source provides much higher magnification (up to 100,000 times) and much better depth of field, unique imaging, and the opportunity to perform elemental analysis and phase identification.

6.6.3 Case Study: PAS (Photo Angle Sensor) in Automobile

- Summary: No ignition of automobile due to the frequent failures of the PAS application IC in field (Fig. 6.25a)
- Electrical test (by curve tracer): Electrical open of Pin #4 (Fig. 6.25b)
- Nondestructive inspection by scanning acoustic microscope, X-Ray radiography

6.6 Failure Analysis

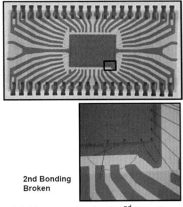
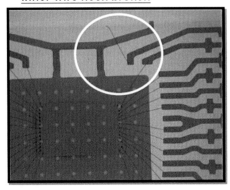

(a) X-ray analysis of 2nd Bonding Broken and Inner wire neck broken (example)

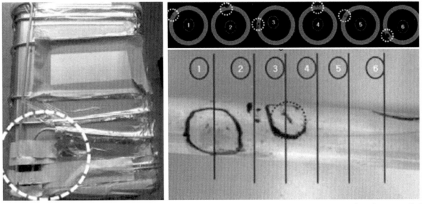

(b) X-Ray Microscopy showing a pitting corrosion on the evaporator tube (example)

Fig. 6.24 Failure analysis using X-ray micrography in product

- SAM : Die paddle(top, bottom), lead frame delamination (Fig. 6.25c)
- X-ray radiography : Package crack (Fig. 6.25d)
- Microscopy analysis (by SEM): Wedge bond open of Pin #4 (Fig. 6.25e).

For delamination in semiconductor, when surface mount device is mounted, because the whole package is exposed to high temperature and humidity, there are problems such as delamination of resin from frame materials or absorbed moisture inside package vapor blasts, and resulting in package deformation or popcorning crack (Fig. 6.26).

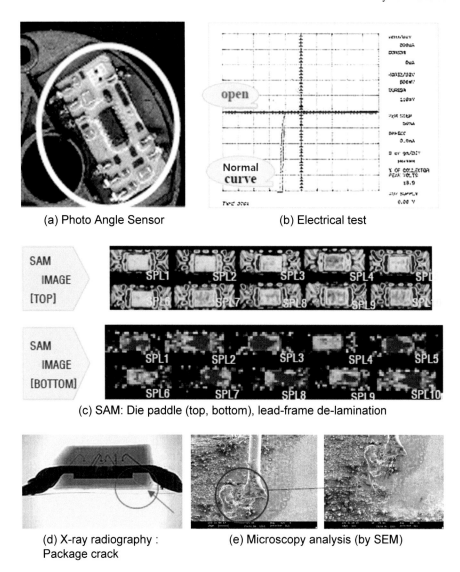

Fig. 6.25 Failure analysis: PAS application IC in automobile

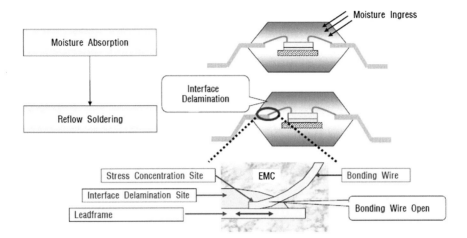

Fig. 6.26 Failure mechanism: delamination in semiconductor

6.6.4 Fracture Faces of Product Subjected to a Variety of Loads in Fields

Fatigue failure can be recognized from the specimen fracture surface with the different growth zones and the major physical features: (1) region of slow crack growth is usually evident in the form of a "clamshell" concentric around the location of the initial flaw, (2) clamshell region often contains concentric "beach marks" at which the crack may become large enough to satisfy the energy or stress intensity criteria for rapid propagation, (3) final phase produces the granular rough surface before final brittle fracture.

For example, the suction reed valves open and close to allow refrigerant to flow into the compressor during the intake cycle of the piston. Due to repetitive stresses, the suction reed valves of domestic refrigerator compressors used in the field were cracking and fracturing, leading to failure of the valve. From SEM microscopy, the fracture started in the void of the suction reed valve and propagated to the end (Fig. 6.27).

The fracture face of a fatigue failure shows both the load type (bending, tension, torsion or a combination) and the magnitude of the load. To understand the type of load, look at the direction of crack propagation. It is always going to be perpendicular to the plane of maximum stress. The following examples reflect the fracture paths on accordance with a variety of loads.

Figure 6.28 describes the reversed torsional fatigue failure of splined shaft from a differential drive gear. The mating halves of the fracture reveal how two separate cracks initiated in a circumferential recess adjacent to the end of the splines and began to propagate into the cross section following helical paths. Because the cycles of twisting forces acted in opposite directions, each crack follows opposing helices which progressively reduced the effective cross-sectional area and,

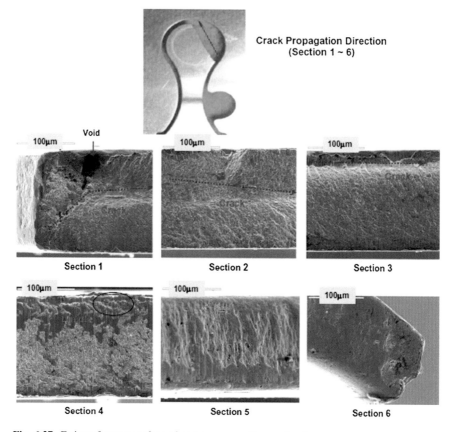

Fig. 6.27 Fatigue fracture surface of compressor suction reed valve

Fig. 6.28 Fatigue failure of splined shaft due to reversed torsional loads

consequently, increased the levels of cyclic stresses from the same applied loads. Shortly before the shaft finally broke bending forces initiated a third crack at the opposite side of the shaft and this had begun to propagate as a plane fracture at 90° to the shaft axis until the splined end finally broke away.

Torsional fatigue is involved in 10–25% of rotating equipment failures. Torsion fatigue failures can identify them as the fracture oriented 45° to the shaft centerline. The fracture face typically has one or more origins, a fatigue zone with progression lines and an instantaneous zone. A large fatigue zone and small instantaneous zone mean the fatigue load was small. A small fatigue zone and large instantaneous zone mean the fatigue load was high.

Torsional fatigue fractures frequently occur in a shaft that is inside a hub or coupling. These fractures usually start at the bottom of a keyway and progress around the shaft's circumference. In Fig. 6.28, the fracture travels around the shaft, climbing toward the surface so the outer part of the shaft looks like it was peeled away. The fracture surface has characteristics of a fatigue fracture: one or more origins, ratchet marks, and a fatigue zone with progression lines. The shaft fragment is usually held in place by the coupling or hub, so there is typically a very small or no instantaneous zone.

A shaft fracture may have both torsion and bending fatigue forces. When this occurs, the orientation of the fracture face may vary from 45° to 90° with respect to the shaft centerline. As the fracture is closer to 90°, the shaft combines dominant bending with torsion. The fracture angle therefore offers key evidence as follows:

- Closer to 90°, it is a dominant bending force.
- Midway between 45° and 90°, it is a combination of torsion and bending forces.
- Closer to 45°, it is a dominant torsion force.

Evidence of torsional fatigue also may be found on gear and coupling teeth. Most equipment runs in one direction, so wear is expected on one side of a gear or coupling teeth. Wear on both sides of a gear or coupling teeth that rotate in one direction is an indication of varying torsional force. When coupling alignment is good and wear occurs uniformly on both sides of all coupling teeth, it usually indicates torsional vibration. Alignment quality can be verified from vibration spectra and phase readings. An absence of 2× running speed spectral peaks and uniform phase across the coupling occurs when the alignment is good.

References

1. Timoshenko SP (1953) History of strength of materials. McGraw-Hill Book Co., New York
2. Anderson TL (1991) Fracture mechanics: fundamentals and applications. CRC Press, Boca Raton
3. Griffith AA (1921) The phenomena of rupture and flow in solids. Philos Trans Roy Soc London A 221:163–198
4. Neugebauer GH (1943) Stress concentration factors and their effect in design. Prod Eng NY A 14:82–87

5. Paris PC, Gomez MP, Anderson WE (1961) A rational analytic theory of fatigue. Trend Eng 13:9–14
6. Wöhler A (1870) Über die Festigkeitsversuche mit Eisen und Stahl. Zeitschrift für Bauwesen 20:73–106
7. ASME (1965) Mechanical reliability concepts. In: ASME design engineering conference. ASME, New York

Chapter 7
Parametric Accelerated Life Testing in Mechanical/Civil System

Abstract In this chapter, as new quantitative methodology in reliability-embedded developing process, parametric Accelerated Life Testing (ALT) will be discussed. Engineer in the design process has final goal to find the problematic part of product and achieve the reliability target. However, it has pending questions—the testing time and sample size. If fewer or limited samples are selected, the statistical assessment for product reliability becomes more uncertain. If a sufficient quantity of parts for more accurate result is tested, the cost and time will demand considerably. It therefore is reasonable to proceed the accelerated life testing. The parametric ALT is shortly carried out until a certain number of failed components have been reached under accelerated conditions. As a reliability methodology, parametric ALT has to derive the sample size equation with accelerated factors. The accelerated factors could be found in analyzing the load conditions of real dynamics system. Typical failure mechanisms in mechanical system are fatigue and fracture. To achieve the reliability target of product, parametric ALT should search out the missing design parameters to robustly withstand the loads in product lifetime.

Keywords Parametric accelerated life testing (ALT) · Loading conditions · Sample size equation · Accelerated factor (AF) · Loading conditions

7.1 Introduction

Reliability describes the ability of a system or module to function under stated conditions for a specified period of time [1]. Reliability is often illustrated in a diagram called "the bathtub curve" shown as the top curve in Fig. 7.1. The first part of the curve, called the "infant mortality period," represents the introduction of the product in the market. In this period, there is a decreasing rate of failure. It is then followed by what is usually called the "normal" life period with a low but consistent failure rate. It then ends with a sharp increase in failures as the product reaches the end of its useful life. If product in the mechanical/civil systems were to exhibit the failure profile in the bathtub curve with a large number of failures in the early life of product, it would be difficult for the system to be successful in the marketplace. Improving the reliability of a system through systematic testing should reduce its failure rate from the traditional failure rate typified by the bathtub curve to the failure rate represented by a flat, straight line with the shape parameter β in Fig. 7.1. With the second curve, there are low failure rates throughout the lifetime of the system or component until reaching the end of its useful life that the failure rate begins to increase.

The product reliability function can be quantified from the expected product lifetime L_B and failure rate λ in Fig. 7.1 as follows:

$$R(L_B) = 1 - F(L_B) = e^{-\lambda L_B} \cong 1 - \lambda L_{BX}. \tag{7.1}$$

In a practical sense, this proportionality is applicable below about 20% of cumulative failure rate [2]. Improving the design of a mechanical system to increase its reliability can be achieved by quantifying the targeted product lifetime L_B and failure rate λ by finding the appropriate control parameters affecting reliability and then modifying the design with the results from parametric accelerated life testing.

7.2 Reliability Design in Mechanical System

As setting an overall parametric accelerated life testing plan, reliability design of the product can be achieved by getting the targeted reliability of product—lifetime L_B and failure rate λ after finding the missing control parameters and modifying the defective configuration of structures.

The product (or module) with the modified design might meet the assigned reliability target. As product (or module) carries out test for significantly longer time, the parametric ALTs might obtain the missing design parameters in the design phase of the mechanical system. Under consumer usage conditions, these new reliability methodologies in the reliability-embedded design process will provide

Fig. 7.1 Bathtub curve and straight line with slope β toward the end of the life of the product

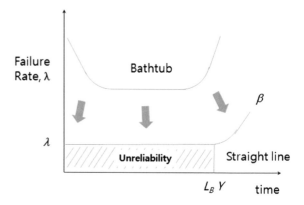

7.2 Reliability Design in Mechanical System

the reliability quantitative (RQ) test specifications of a mechanical structure that conforms to the reliability target.

As shown in Fig. 7.2, a product can consist of several different modules. For example, automobiles consist of modules, such as the engine, transmission, drive, electrical, and body parts. The product lifetime L_B and failure rate λ_s with multi-modules should be determined for each module. For example, suppose that there were no initial failures in a product, the product lifetime could be represented by the product lifetime for module #3 in Fig. 7.2. The cumulative failure rate of the product over its lifetime would be the sum of the failure rate of each module as shown in Fig. 7.2b. One core module #3 will seriously damage the reliability of the whole product and determine the product lifetime. If the product lifetime was given by Y and the total failure rate was X, the yearly failure rate can be calculated by dividing total failure rate X by product lifetime Y. The product reliability may be given as reliability $(1 - X * 0.01)$ with a yearly failure rate of X/Y and L_{BX} Y years.

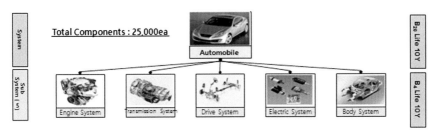

(a) Breakdown of Automobile with multi-modules

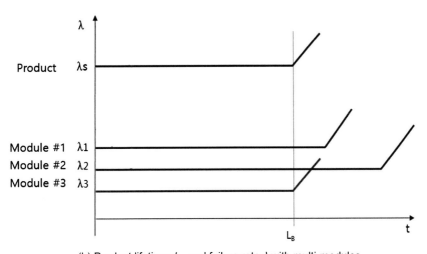

(b) Product lifetime L_B and failure rate λ with multi-modules

Fig. 7.2 Product lifetime L_B and failure rate λ_s with multi-modules

Based on failure data from the field, the parametric accelerated life testing plan of the product can be established for a newly designed module and any modified module. Table 7.1 shows the parametric ALT for several modules. For module D, a modified module, the yearly failure rate was 0.2%/year and L_{Bx} life was 6 years from the field data. Because this was a modified design, the expected failure rate was 0.4%/year and the expected L_{Bx} life was 3.0. To increase the targeted product life, the lifetime of the new design was targeted to be L_{Bx} ($x = 1.2$) 12 years with a yearly failure rate of 0.1%. The product reliability might be determined by summing the failure rates of each module and lifetimes of each module. The product reliability is targeted to be over a yearly failure rate of 1.1% and L_{Bx} ($x = 13.2$) 12 years (Table 7.1).

In targeting the reliability of the new module where there was no field reliability data, the data for similar modules are often used as a reference. If there has been major redesign of the module, the failure rate in the field may be expected to be higher. Thus, the predicted failure rates will depend on the following factors:

1. How well the new design maintains a similar structure to the prior design,
2. For each new module, new manufacturers are assumed to supply parts for the product,
3. Magnitude of the loads compared to the prior design, and
4. How much technological change and additional functions are incorporated into the new design.

Table 7.1 Overall parametric ALT plan of product

No	Reliability Modules	Market data		Expected design			Targeted design		
		Yearly failure rate (%/year)	B_x life, year	Yearly failure rate (%/year)		B_x life (year)	Yearly failure rate (%/year)	B_x life (year)	
1	Module A	0.34	5.3	New	×5	1.70	1.1	0.15	12 ($x = 1.8$)
2	Module B	0.35	5.1	Given	×1	0.35	5.1	0.15	12 ($x = 1.8$)
3	Module C	0.25	4.8	Modified motor	×2	0.50	2.4	0.10	12 ($x = 1.2$)
4	Module D	0.20	6.0	Modified	×2	0.40	3.0	0.10	12 ($x = 1.2$)
5	Module E	0.15	8.0	Given	×1	0.15	8.0	0.1	12 ($x = 1.2$)
6	Others	0.50	12.0	Given	×1	0.50	12.0	0.5	12 ($x = 6.0$)
Total	R-set	1.79	7.4	–	–	3.60	3.7	1.10	12 ($x = 13.2$)

So for Module A, the expected failure rate was 1.7%/year and its expected lifetime was 1.1 years because there was no field data on the reliability of the new design. The reliability of the new design was targeted to be over L_{Bx} ($x = 1.8$) 12 years with a yearly failure rate of 0.15%. To meet the expected product lifetime, the parametric ALT should help identify design parameters that could affect the product reliability.

7.3 Reliability Block Diagram and Its Connection in Product

As shown in Fig. 7.3, the reliability block diagram is a graphical method that describes how system and main module connected in product. The configurations of complicated system such as automobile can be generated from the series or parallel connections between modules. In a reliability block diagram, components are symbolized by rectangular blocks, which are connected by straight lines according to their logic relationships. Depending on the purpose of system analysis, a block may represent a lowest-level component, module, subsystem, and system. It is treated as a block box for which the physical details may not need to be known. The reliability of the object that a block represents is the only input that connects system reliability evaluation.

In constructing a reliability block diagram, physical configurations in series or parallel do not indicate the same logic relations from a standpoint of reliability. A system is said to be a series system if the failure of one or more modules within the system result in failure of the entire system. A variety of mechanical products are the serial system at two hierarchical levels that consist of multiple modules. For example, in an automobile engine, six cylinders are in series because the engine is said to have failed if one or more cylinders connected in parallel mechanically are

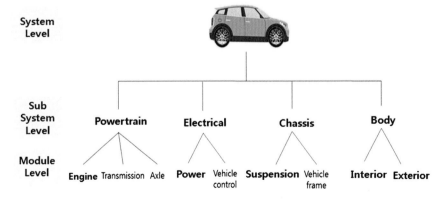

Fig. 7.3 Automobile that consists of multi-modules connected serially

failed. In the same manner automobile is serially connected in power train, electrical and control system, chassis, and body.

Suppose that a mechanical series system like automobile consists of n mutually independent modules. Mutual independence implies that the failure of one module does not affect the life of other modules. By definition, successful operation of a system requires all components to be functional. From probability theory, the system reliability is

$$R = Pr(E) = Pr(E_1 \cdot E_2 \cdots E_n), \qquad (7.2)$$

where E_i is the event that module i is operational, E is the event that the system is operational, and R is the system reliability.

Because of the independence assumption, this becomes

$$R = Pr(E) = Pr(E_1) \cdot Pr(E_2) \cdots Pr(E_n) = \prod_{i=1}^{n} R_i, \qquad (7.3)$$

where R_i is the reliability of module i.

Let us consider a simple case where the times to failure of n modules in a system are modeled with the exponential distribution. The exponent reliability function for module i is the exponential reliability function for component i: $R_i(t) = \exp(-\lambda t)$, where λ_i is the failure rate of component i. Then from (7.3), the system reliability can be written as

$$R(t) = \exp\left(-t \sum_{i=1}^{n} \lambda_i\right) = \exp(-\lambda t), \qquad (7.4)$$

where λ is the failure rate of the system and $\lambda = \sum_{i=1}^{n} \lambda_i$.

7.4 Reliability Allocation of Product

7.4.1 Introduction

If the system reliability target is setting in the product planning, it will sequentially be allocated to individual subsystem, module, and components at the stage of the product design. When each module achieves the allocated reliability, the overall system reliability target can be attained. Reliability allocation is an important step in the new reliability testing design process. The benefits of reliability allocation can be summarized as follows:

- Reliability allocation defines a reliability target for each module. The product has a number of module or subsystem, which are manufactured by suppliers or

internal departments. It is important that company share all related parties with the reliability target before delivering the end product to customer.
- Quantitative reliability targets for modules encourage responsible parties to improve current reliability through use of reliability techniques.
- Mandatory reliability requirements are closely connected with engineering activities aimed at meeting other customer expectations in the product design process.
- Reliability allocation drives a deep understanding of product hierarchical structure. The process may lead to identify the part of design weakness and subsequently improve it.
- As a result, reliability allocation can work on input of other reliability tasks. For example, reliability assigned to a module will be used to design reliability verification.

Reliability allocation is fundamentally a repetitive process. It is conducted in the early design stage to support concept design when available information is restricted. As the design process proceeds, the overall reliability target might be reallocated to reduce the cost of achieving the reliability goal. The allocation process may be invoked by the failure of one or more modules to attain the assigned reliability due to technological limitations. The process is also repeated whenever a major design change takes place.

7.4.2 Reliability Allocation of the Product

Because some parts are assigned to extremely high-reliability goals, it may be unachievable at all. On the other hand, though there are critical components whose failure causes safety, environmental or legal consequences, it will be allocated to low-reliability targets. It is important to establish some criteria that should be considered in reliability allocation.

The task of reliability allocation is to select part reliability targets, $R_1^*, R_2^*, \ldots, R_n^*$ which satisfy the following equality equation:

$$R_S^* \leq g\left(R_1^*, R_2^*, \ldots, R_n^*\right). \tag{7.5}$$

Mathematically, there are an infinite number of such sets. Clearly, these sets are not equally good, and even some of them are unfeasible. The common criteria are described:

1. Failure possibility. Parts that have a high likelihood of failure previously should be given a low-reliability target because of the intensive effort required to improve the reliability. Conversely, for reliable parts, it is reasonable to assign a high-reliability goal.

2. **Complexity.** The number of constituent parts (or modules) within a subsystem reflects the complexity of the subsystem. A higher complexity leads to a lower reliability. It is similar to the purpose of failure possibility.
3. **Criticality.** The failure of some parts may cause severe effects, including, for example, loss of life and permanent environmental damage. The situation will be severe when such parts have a high likelihood of failure. Apparently, criticality is a product of severity and failure probability, as defined in the FMEA technique described in Chap. 4. If a design cannot eliminate severe failure modes, the parts should have the lowest likelihood of failure. Consequently, high-reliability goals should be assigned to them.
4. **Cost.** Cost is an essential criterion that is a target subject to minimization in the commercial industry. The cost effects for achieving reliability depend on parts. Some parts induce a high cost to improve reliability a little because of the difficulty in design, verification, and production. So it may be beneficial to allocate a higher reliability target to the parts that have less cost effect to enhance reliability.

Though several methods for reliability allocation have been developed, the simplest method here is the equal allocation method. This method can only be applied when the system reliability configuration is in series. The system reliability is calculated by

$$R_S^* = \prod_{i=1}^{n} R_i. \tag{7.6}$$

The allocated reliability for each subsystem is

$$R_i = (R_S^*)^{1/k}. \tag{7.7}$$

7.4.3 Product Breakdown

Typical modern products involved in mechanical system can be outlined as automobile, airplane, domestic appliance, machine tools, agricultural machinery, and heavy construction equipment. They can break down several modules to the individual parts. Based on the market data, the reliability target could be assigned to the product modules as in Table 7.2. The targeted reliability of each module can be quantified as the expected product lifetime L_B and failure rate λ in Eq. (7.1). The reliability testing will be centered on the module of product. For example, if the targeted reliability of refrigerator is allocated as B20 life 5 year, the reliability for engine will be B4 life 5 year.

It is reasonable to carry out the reliability testing per module because test cost for system or component is higher than that of module.

7.4 Reliability Allocation of Product

Table 7.2 Reliability target for mechanical system (part count method)

Level	Quantity	Target	Remark
System	1 system	B20 life 10 year	Refrigerator
Module	5 units	B4 life 10 year	Compressor
Component	500 Components	B0.04 life 10 year	–

7.4.3.1 Automobile

Figure 7.4 shows the hierarchical configuration of an automobile connected serially from system to main modules. It consists of engine, body and main parts, electrical and electronics, interior, power train and chassis, miscellaneous auto parts, air conditioning system (A/C), bearings, hose, and other miscellaneous parts. An automobile is a four-wheeled, self-powered motor designed to run on roads for transport of one to eight people. Each subsystem in automobile is broken down further into multiple lower level subsystems. From a reliability perspective, the automobile is a series system which fails one or more subsystems (or module) in automobile break. The blocks of the automobile in the reliability block diagram represent the first-level subsystems and the second-level modules, where their reliabilities are known. The reliability block diagram of a typical automobile contains over 20,000 blocks, including the parts.

However, reliability design of automobile might focus on the modules. They can easily calculate the module reliability because of the connection of serial system.

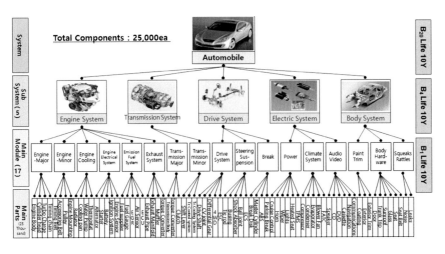

Fig. 7.4 Breakdown of automobile with multi-modules

7.4.3.2 Airplane

The uses for airplanes are constructed with the objectives of recreation, transportation of goods and people, and military. The design and planning process, including safety tests, can last up to 4 years. The design specifications of the aircraft during the design process often is established. When the design has passed through these processes, the company constructs a limited number of prototypes for testing on the ground.

Figure 7.5 shows the hierarchical configuration of a passenger airplane that consists of airframe parts, wings, fuselage, propulsion (engine), aviation controls and instruments, air conditioning system (A/C), bearings, hose, and other miscellaneous airplane parts. From a reliability perspective, the airplane is a series system which fails one or more subsystems (or module) in airplane break. The blocks of the airplane in the reliability block diagram represent the first-level subsystems and the second-level modules, where their reliabilities are known. The reliability block diagram of a typical airplane contains over 1,000,000 blocks, including the parts. Reliability design of airplane will focus on the modules that are serially connected like other mechanical system shown in Fig. 7.5.

7.4.3.3 Domestic Appliance

Domestic appliance is a large machine used for routine housekeeping tasks such as cooking, washing laundry, or food preservation. Examples include refrigerator, air conditioner, washing machine, and cleaner. Major appliances that use electricity or fuel are bigger and not portable. They are often supplied to tenants as part of otherwise unfurnished rental properties. Major appliances may have special

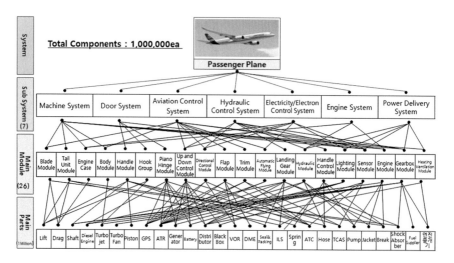

Fig. 7.5 Breakdown of airplane with multi-modules

7.4 Reliability Allocation of Product

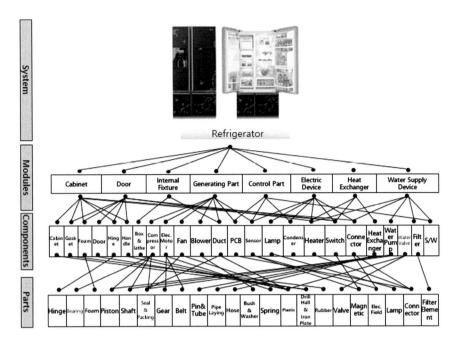

Fig. 7.6 Breakdown of a refrigeration system with multi-modules

electrical connections, connections to gas supplies, or special plumbing and ventilation arrangements that may be permanently connected to the appliance. This limits where they can be placed in a home (Fig. 7.6).

The hierarchical configuration of an appliance consists of cabinet, door, internal fixture (selves, draws), controls and instruments, generating parts (motor or compressor), heat exchanger, water supply device, and other miscellaneous parts. The reliability block diagram of a typical appliance contains over 1000 blocks which is including the parts. Reliability design of domestic appliance will focus on the modules. They can easily calculate the module reliability because of the connection of serial system.

7.4.3.4 Machine Tools

A machine tool is a machine for shaping or machining metal or other rigid materials, usually by cutting, boring, grinding, shearing, or other forms of deformation. Machine tools employ some sort of tool that does the cutting or shaping. All machine tools have some means of constraining the work piece and provide a guided movement of the parts of the machine. Thus the relative movement between the work piece and the cutting tool (which is called the tool path) is controlled or constrained by the machine to at least some extent, rather than being entirely "offhand" or "freehand" (Fig. 7.7).

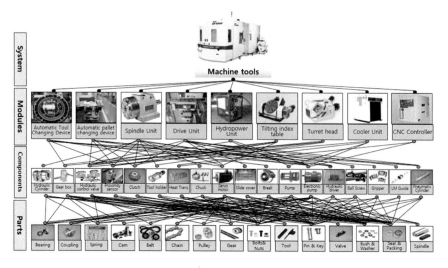

Fig. 7.7 Breakdown of machine tools with multi-modules

The hierarchical configuration of machine tools consists of automatic tool or pallet changing device, spindle unit, drive unit, hydro-power unit, tilting index table, turret head, cooler unit, CNC controller, and other miscellaneous parts. The reliability block diagram of typical machine tools contains over 1000 blocks, including the parts. Reliability design of machine tools will focus on the modules. They can easily calculate the module reliability because of the connection of serial system.

7.4.3.5 Agricultural Machinery and Heavy Construction Equipment

Agricultural machinery such as tractor is used in the operation of an agricultural area or farm. The hierarchical configuration of agricultural machinery such as automobile consists of engine device, power supply unit, hydraulic unit, electric device, linkage, PTO driving unit, and other miscellaneous parts (see Fig. 7.8). The reliability block diagram of a typical appliance contains over 4000 blocks, including the parts.

The hierarchical configuration of a construction machine such as excavator consists of engine device, electric device, track system, upper appearance system, driving system, main control valve unit, hydraulic operation machine system, cooling system, and other miscellaneous parts. The reliability block diagram of a typical appliance contains over 5000 blocks including the parts (see Fig. 7.9).

Heavy equipment refers to heavy-duty vehicles, specially designed for carrying out construction tasks, most frequent ones operating earthwork. They are also

7.4 Reliability Allocation of Product

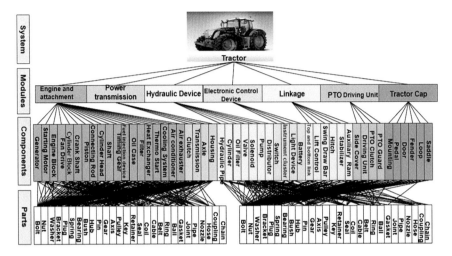

Fig. 7.8 Breakdown of tractor with multi-modules

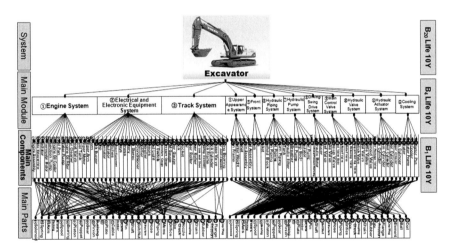

Fig. 7.9 Breakdown of excavator with multi-modules

known as heavy machines, heavy trucks, construction equipment, heavy vehicles, or heavy hydraulics. They usually comprise five equipment systems: engine, traction, structure, power train, control, and information. Some equipment frequently uses hydraulic drives as a primary source of motion. Reliability design of agricultural machinery and heavy construction equipment will focus on the modules shown in Fig. 7.9. They can easily calculate the module reliability because of the connection of serial system.

7.5 Failure Mechanics, Design, and Reliability Testing

The failure mechanics of product can be described as two factors: (1) the stress on the structure, (2) the structure materials. The failure mechanisms can be characterized by either loads (or stress)-related failure or structural (or materials)-related failure.

If there is void in the product structure, the structured will facture as shown in Fig. 7.10. On the other hands, if structure has enough safety margins for load, the structure will deteriorate little by little and facture near product lifetime. In field two cases—stress and material happen complexly. The typical failure mechanisms of mechanical system are fracture and fatigue. From bathtub curve, this region would be described as the constant failure rate that often receives repetitive random stress. As the repeated load is applied at the stress raisers such as shoulder filet, the structure that damage is accumulated will crack. After repetitive stresses, the system will fracture suddenly.

Failure of mechanical/civil systems can happen when the system structures yield at the strength of materials by the applied loads. The load could be higher than the system was designed for. On the other hand, the material could be insufficient to handle repetitive loads to which it is subjected. Consequently, failure occurs when the stress is greater than the material strength or when the material cannot withstand the loads. The product engineer would want to move the void in the structure to a location away from where the stress is applied. This is a design concept.

A product engineer should seek to redesign the structure to either (1) move the loads or (2) change the material type and design shape to withstand the load. The failure site of the product structure could be found when the failed products are taken apart in the field or after the failed samples of a parametric ALT. The engineer should identify the failure by experiment using the reliability testing (or parametric ALT) before launching new product. This failure phenomenon, design, and reliability testing might be applicable to both mechanical and electronic products because the

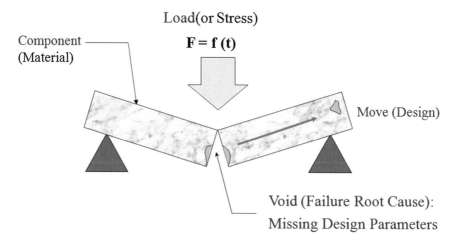

Fig. 7.10 Failure mechanics and mechanical system design

7.5 Failure Mechanics, Design, and Reliability Testing

electric products are typically housed in mechanical systems. So it is a critical process to search out void such as the stress raisers that have the missing design parameters by FEA or experimentally using the reliability testing (or parametric ALT).

With the advent of Finite Element Analysis (FEA) tools, design failure such fatigue can now be assessed in a virtual environment. Although FEA fatigue assessments do not completely replace fatigue testing, they will find the detailed or optimal design in the structure of new product. However, as the system goes from preliminary design to an optimized design, product might have defects. If modules have a problem due to an improper design, the module will determine the lifetime of the product (Fig. 7.11 on the left side).

Even if a detailed product design is optimized using tools such as FEA, product may have still design flaws that do not become evident until the system is in the field. In the field, the system may be subjected to loads that could be very different than what was envisioned in the original design. Field data can be used by both the design engineer and test engineer to develop appropriate parameters for accelerated life tests that help validate the expected reliability of the design. Figure 7.11 shows how for any system, design engineering needs to be effectively connected with test engineering to achieve the reliability target of a system, subsystem, or module.

As described in Sect. 7.4, the mechanical products such as appliance, car, and aircraft consist of a multiple of modules. These modules can be put together in a system and have an input and output similar to what is shown in Fig. 7.12. They also have their own (intended) function just as a vapor compressor cycle that generates the cold air.

In the field, if a mechanical/civil module functions improperly, consumers would request the module replaced. Reliability engineers often do not have a clear understanding of the way that a consumer used the product or the usage patterns

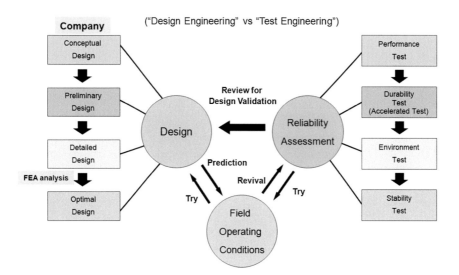

Fig. 7.11 Optimal design and reliability assessment

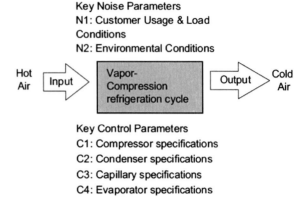

Fig. 7.12 Typical robust design schematic (example: refrigerator)

that could have contributed to the failure of the product. If the field usage conditions are fully understood, they could be reproduced in the laboratory testing to be identical to those of the failure in field. In the design of the product, it is important to understand potential usage patterns and take corrective actions before launching a product.

Under a variety of environmental and operational conditions, reliability engineers search for potential failure modes by testing in the laboratory. They will determine the failure mechanisms from the failed samples. They can create an action plan to reduce the chance of failure. However, it may not be easy to identify all failure modes attributable to the improper design because, in mechanical/civil systems, the failure modes come from repetitive stresses which may not be captured in initial testing.

Consequently, modules with specific functions need to be robustly designed to withstand a variety of loads. In determining product lifetime, the robust design in module determines the control factor (or design parameters) to endure the noise factor (or stress) and properly work the system, which has the reliability target—part failure rate λ and lifetime L_B. Such reliability targeting is known to be conventionally achieved through the Taguchi methods (SDE) and the statistical design of experiment [3–8].

Taguchi methods, known to robust designs, use the loss function which quantifies the amount of loss based on deviation from the target performance. It puts a design factor in an optimal location where using cost function random "noise" factors is less likely to hurt the design and it helps determine the best control factors (or design parameters). However, for an uncomplicated mechanical/civil structure, such as a beam, Taguchi methods should take into account a considerable number of design parameters. In the design process it is not possible to consider the whole range of the physical, chemical, and the mathematical conditions that could affect the design.

Another experimental methodology, new parametric ALT methods for reliability quantitative test specifications (RQ), should be introduced so that the product can withstand a variety of repetitive loads and determine the critical design parameters affecting reliability. Parametric ALT can also be used to predict product reliability —lifetime L_B and failure rate λ. The new parametric ALT discussed in the next section has a sample size formulation that enables an engineer to determine the design parameters and achieve the targeted product reliability—lifetime L_B and failure rate λ [9–21].

7.6 Parametric Accelerated Life Testing

Parametric accelerated life testing uses the sample size equation with Acceleration Factor (AF). It is also a process that helps designers to find the optimal (or missing) design parameters. If the reliability target of module in product is allocated, module should accomplish Reliability Quantitative (RQ) test specifications by obtaining the sample size equation. It can help them better estimate expected lifetime L_B, failure rate of module λ, and finally determine whether the overall product reliability is achieved.

The shape parameter β in the sample size equation is calculated from a Weibull distribution chart tested. From the sample size equation, the durability target h^* is determined by the targeted lifetime L_B, acceleration factor AF, and the actual testing time h_a. So it is important to derive the sample size equation with the whole parameters—lifetime L_B, acceleration factor AF, the actual testing time h_a, and the allowed number of failures.

Under the expected physical and chemical conditions that the product is expected to experience, it is essential to derive the acceleration factor from a life–stress model, and determine the dominant failure mechanism for the product. A grasp of Physical of Failure (PoF) also is required to understand the failure mechanism. For example, fatigue or fracture due to repetitive stresses is the common mechanism for failure in mechanical/civil system.

Reliability engineers must also determine how the stresses (or loads) act on the system structure, which help categorize the potential failure mechanisms under the range in environmental and operational conditions. Engineers need to develop a testing plan with appropriate accelerated load conditions to determine the dominant failure mechanisms affecting product lifetime. At the same time, they also must include other failure mechanisms, such as overstress and wear-out stress. The failure mechanisms in the accelerated life testing should be identical to that under normal conditions experienced in the field. In the Weibull distribution, the shape parameter for accelerated conditions should match those under normal field conditions.

Developing a parametric ALT for reliability quantitative test specifications involves three key steps:

1. By creating a life–stress model, determine the acceleration factor under severe conditions.

2. Assuming an initial shape parameter that is implied by the intensity distribution of wear failure in the Weibull distribution, derive the necessary sample size to carry out the lifetime (or reliability) target and calculate the testing periods (or Reliability Quantitative (RQ) test specifications) equivalent to the allocated reliability target.
3. With sample size equation, carry out testing for extended test periods to help determine the actual shape parameter in the Weibull distribution.

Degradation by loads is a fundamental phenomenon to all products that this is described as entropy of isolated systems which will tend to increase with time—the second law of thermodynamics. For instance, the critical parameters such as strength will degrade with time. In order to understand the useful lifetime of the part, it is important to be able to model how critically important product parameters degrade with time.

7.6.1 Acceleration Factor (AF)

Reliability concerns arise when some critically important mechanical/civil strength due to stress degrades with time. Let S represent a critically important part parameter like strength and let us assume that S change monotonically and relatively slowly over the lifetime of the part. A Taylor expansion about $t = 0$ produces the Maclaurin series like Eq. 7.8:

$$S(t) = S_{t=0} + \left(\frac{\partial S}{\partial t}\right)_{t=0} t + \frac{1}{2}\left(\frac{\partial S}{\partial t}\right)_{t=0} t^2 + \cdots \qquad (7.8)$$

It will be assumed that the higher order terms in the expansion can be approximately by simply introducing a power-law exponent m and writing the above expansion in a shortened form:

$$S = S_0[1 \pm A_0(t)^m], \qquad (7.9)$$

where A_0 is a part-dependent constant.

The power-law model is one of the most widely used forms for time-dependent degradation. For convenience of illustration, let us assume that the critical parameter S is decreasing with time and $A_0 = 1$. Equation (7.9) reduces to

$$1 - \frac{S}{S_0} = (t)^m. \qquad (7.10)$$

For $m = 1$, one will expect the linear degradation. On the other hand, for $m > 1$, the degradation will increase strongly with time and eventually lead to a catastrophic condition.

7.6 Parametric Accelerated Life Testing

In reliability engineering, the development of the acceleration factor is fundamental importance to the theory of accelerated life testing. The acceleration factor speeds up the degradation of product and permits one to take time-to-failure data very rapidly under accelerated stress conditions. And it can extrapolate the accelerated time-to-failure results into the future for a given set of operational conditions. The acceleration factor must be modeled using the time-to-failure (TF) models. The acceleration factor is defined as the ratio of the expected time-to-failure under normal operating conditions to the time-to-failure under some set of accelerated stress conditions:

$$AF = \frac{(TF)_{operation}}{(TF)_{stress}}. \tag{7.11}$$

Since the TF under normal operation may take many years to occur, experimental determination of the acceleration factor is impractical. However, if one has proper time-to-failure models, one can develop the acceleration factor from the TF models.

For solid-state diffusion of impurities in silicon, the junction equation J might be expressed as

$$J = [aC(x-a)] \cdot \exp\left[-\frac{q}{kT}\left(w - \frac{1}{2}a\xi\right)\right] \cdot v$$

[Density/Area] · [Jump Probability] · [Jump Frequency]

$$= -[a^2 v e^{-qw/kT}] \cdot \cosh\frac{qa\xi}{2kT}\frac{\partial C}{\partial x} + [2ave^{-qw/kT}]C\sinh\frac{qa\xi}{2kT}$$

$$= \Phi(x, t, T)\sinh(a\xi)\exp\left(-\frac{Q}{kT}\right) \tag{7.12}$$

$$= A\sinh(a\xi)\exp\left(-\frac{Q}{kT}\right),$$

where Q is the energy, a is the coefficients, and A is the constants.

Reaction process that is dependent to speed might be expressed as

$$K = K^+ - K^- = a\frac{kT}{h}e^{-\frac{\Delta E - aS}{kT}} - a\frac{kT}{h}e^{-\frac{\Delta E + aS}{kT}}$$

$$= a\frac{kT}{h}e^{-\frac{\Delta E}{kT}}\sinh\left(\frac{aS}{kT}\right), \tag{7.13}$$

where k is the Boltzmann's Constant, T is the absolute temperature, S is the stress, and A is the constants.

So the reaction rate K can be summarized as

$$K = B \sinh(aS) \exp\left(-\frac{E_a}{kT}\right), \qquad (7.14)$$

where E_a is the activation energy and B is the constants.

If the reaction rate in Eq. (7.14) and the junction Eq. (7.12) takes an inverse number, the generalized stress model can be obtained like McPherson's derivation [22]:

$$\text{TF} = A[\sinh(aS)]^{-1} \exp\left(\frac{E_a}{kT}\right). \qquad (7.15)$$

Because this life–stress model, equation was derived from a model of micro-depletion (void) in the failure domain, and it should be relevant to general failure prediction regardless of whether it is a mechanical, civil, or electronic system. Thus, the fatigue in a mechanical/civil system, coil degradation in a motor, bond-pad corrosion in an IC, etc., can be captured by Eq. (7.15).

The range of the hyperbolic sine stress term $[\sinh(aS)]$ in Eq. (7.12) is increasing the stress as follows: (1) initially (S) in low effect, (2) $(S)^n$ in medium effect, and (3) $(e^{aS})^n$ in high effect initially linearly increasing. Accelerated testing usually happens in the medium stress range (see Fig. 7.13).

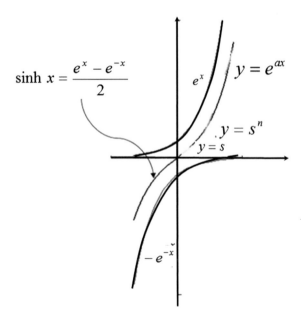

Fig. 7.13 Properties of the hyperbolic sine stress term $[\sinh(aS)]$

7.6 Parametric Accelerated Life Testing

Thus, time-to-failure in the level of medium stress can then be described as

$$\text{TF} = A(S)^{-n} \exp\left(\frac{E_a}{kT}\right). \tag{7.16}$$

The internal (or external) stress in a product is difficult to quantify and use in accelerated testing. It is necessary to modify Eq. (7.16) into a more applicable form. The power (or energy flow) in a physical system can generally be expressed as efforts and flows (Table 7.3). Thus, stresses in mechanical/civil or electrical systems may come from the efforts (or loads) like force, torque, pressure, or voltage [23].

For a mechanical/civil system, when replacing stress with effort, the time-to-failure can be modified as

$$\text{TF} = A(S)^{-n} \exp\left(\frac{E_a}{kT}\right) = A(e)^{-\lambda} \exp\left(\frac{E_a}{kT}\right), \tag{7.17}$$

where λ is the power index or damage coefficient.

Because the material strength degrades slowly, it may require long time to test a module until failure occurs. The main hurdles to find wear-induced failures and overstressed failures are the testing time and cost. To solve these issues, the reliability engineer often prefers testing under severe conditions. Due to overstress failures of the module can be easily found with parametric ALT.

The more the accelerated conditions, the shorter the testing time will be. This concept is critical to perform accelerated life tests, but the range of the accelerated life tests will be determined by whether the conditions in the accelerated tests are the same to that in normally found in the field.

The stress–strain curve is a way to visualize behavior of material when it is subjected to load (see Fig. 7.14). A result of stresses in the vertical axis has the corresponding strains along the horizontal axis. Mild steel subjected to loads passes specification limits (proportional limit), operating limits (elastic limit), yield point, and ultimate stress point into fracture (destruct limit). In accelerated testing, the appropriate accelerated stress levels (S_1 or e_1) will typically fall outside the specification limits but inside the operating limits.

In accelerated life tests, when a module has been tested for a number of hours under the accelerated stressed condition, one wants to know the equivalent operation time at the normal stress condition. The equivalent operation time is obtained

Table 7.3 Energy flow in the multi-port physical system

Modules	Effort, $e(t)$	Flow, $f(t)$
Mechanical translation	Force, $F(t)$	Velocity, $V(t)$
Mechanical rotation	Torque, $\tau(t)$	Angular velocity, $\omega(t)$
Compressor, pump	Pressure difference, $\Delta P(t)$	Volume flow rate, $Q(t)$
Electric	Voltage, $V(t)$	Current, $i(t)$

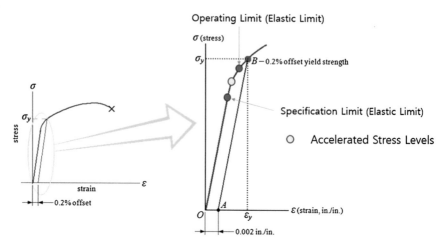

Fig. 7.14 Strain–stress curve in mild steel

from the multiplication of the inverse of acceleration factor and normal (or actual) operation time.

From the time-to-failure in Eq. (7.17), the acceleration factor can be defined as the ratio between the proper accelerated stress levels and normal stress levels. The acceleration factor (AF) also can be modified to include the effort concepts:

$$\text{AF} = \left(\frac{S_1}{S_0}\right)^n \left[\frac{E_a}{k}\left(\frac{1}{T_0} - \frac{1}{T_1}\right)\right] = \left(\frac{e_1}{e_0}\right)^\lambda \left[\frac{E_a}{k}\left(\frac{1}{T_0} - \frac{1}{T_1}\right)\right], \quad (7.18)$$

where n is the stress dependence and λ is the cumulative damage exponent.

It is very important to note that the acceleration factor is very special, in which the acceleration factor is the independent coefficient A. This means that even though the time-to-failure TF must be expressed as a distribution of time-to-failure, the acceleration factor is unique. AF depends on the kinetic value (λ, E_a) and not on part-to-part variation.

The first term is the outside effort (or load) and the second is the internal energy in Eq. (7.18). Under severe conditions, the outside higher load drops the energy barrier and the accelerated (or high) temperature activates the material elements. In the end, the material degrades and fails. The equation has two parameters which are temperature and effort. Using a three-level test under accelerated conditions, these parameters can be obtained, and the quantified value, *activation energy* (eV), is called the reaction rate due to temperature rises.

Under severe conditions, the duty effect with repetitive stress (or load) involves the on/off cycles, which shortens module lifetime [24]. The equation needed to determine the sample size for the parametric ALT is discussed in the next section.

7.6.2 Derivation of General Sample Size Equation

Due to the cost and time limit, it is difficult to test large samples for reliability testing of product. If the fewer components are tested, the greater the confidence interval is, the results of a statistical analysis will become more uncertain. For a more precise result it is necessary that enough samples are tested. This, however, can increase the time and cost (or effort) involved in testing. Thus, it is important to develop the sample size equation with acceleration factor in Eq. (7.19), which allows the core testing methods for securing reliability information as inexpensively as possible in the reliability-embedded product-developing process.

For the lifetime assessment, the confidence levels are necessary because it is not possible to gather the lifetimes of several sample sizes. In statistics, the failure behavior of the limited sample may strongly deviate from the actual failure behavior of the population itself. The core concept in statistics offers a further help through the confidence levels, which can specify the confidence of the test results and estimate the failure behavior of the population.

In statistical test planning the first step involves determining how the sample size should be extracted of the inspection lots or population. The test samples are chosen randomly for a representative test sample. The sample size is connected with the confidence levels and the statistical range of the measured failure values. Another important point establishes a suitable test strategy—complete tests, incomplete (censored) test, and accelerated testing for shortening times.

The best statistical case is a complete test that all test samples of the population are subjected to a lifetime test. This means that the test is run until the last element has failed. Thus, failure times for all elements are available for further assessment. However, people should remember why the lifetime testing in company is completed. That is, for new product, the missing design parameters are found before market launching.

In order to reduce the time and effort involved in a lifetime testing, it is reasonable to carry out censored tests or the accelerated testing. The tests are carried out until a certain predetermined lifetime or until a certain number of failed components have been reached with accelerated condition.

If fewer or limited parts are censored, the statistical assessment becomes more uncertain. If more accurate result is required, a sufficient quantity of parts is tested. In this case the cost and time will be demanded. Thus, to save the testing time, parametric accelerated life testing in mechanical/civil system has to be developed.

From various developed methods to determine sample size, the Weibayes analysis is well-known and widely accepted method. However, its mathematical complexity makes it difficult to apply it directly to determined sample size. Failures ($r \geq 1$) need to be distinguished from no failure ($r = 0$) cases. Hence, it is necessary to develop a simplified sample size equation from the Weibayes analysis.

The Cumulative Distribution Function (CDF) in Weibull can be expressed as

$$F(t) = 1 - e^{-\left(\frac{t}{\eta}\right)^{\beta}}. \tag{7.19}$$

The Weibull reliability function, $R(t)$, is expressed as

$$R(t) = e^{-\left(\frac{t}{\eta}\right)^{\beta}}. \tag{7.20}$$

In statistics, Maximum Likelihood Estimation (MLE) is a method of estimating the parameters of a statistical model—some unknown mean and variance which are given to a data set. Maximum likelihood selects the set of values of the model parameters that maximizes the likelihood function. The characteristic life η_{MLE} from the maximum likelihood estimation can be derived as

$$\eta_{MLE}^{\beta} = \sum_{i=1}^{n} \frac{t_i^{\beta}}{r}. \tag{7.21}$$

If the confidence level is $100(1 - \alpha)$ and the number of failure is $r \geq 1$, the characteristic life, η_{α}, would be estimated from Eq. (7.20):

$$\eta_{\alpha}^{\beta} = \frac{2r}{\chi_{\alpha}^{2}(2r+2)} \cdot \eta_{MLE}^{\beta} = \frac{2}{\chi_{\alpha}^{2}(2r+2)} \cdot \sum_{i=1}^{n} t_i^{\beta} \quad \text{for } r \geq 1. \tag{7.22}$$

Presuming there is no failures, p value is α and $\ln(1/\alpha)$ is mathematically equivalent to chi-squared value, $\frac{\chi_{\alpha}^{2}(2)}{2}$. The characteristic life, η_{α}, would be represented as

$$\eta_{\alpha}^{\beta} = = \frac{2}{\chi_{\alpha}^{2}(2)} \cdot \sum_{i=1}^{n} t_i^{\beta} = \frac{1}{\ln \frac{1}{\alpha}} \cdot \sum_{i=1}^{n} t_i^{\beta}, \quad \text{for } r = 0. \tag{7.23}$$

Thus, Eq. (7.22) is established for all cases $r \geq 0$ and can be redefined as follows:

$$\eta_{\alpha}^{\beta} = \frac{2}{\chi_{\alpha}^{2}(2r+2)} \cdot \sum_{i=1}^{n} t_i^{\beta}, \quad \text{for } r \geq 0. \tag{7.24}$$

To evaluate the Weibull reliability function in Eq. (7.24), the characteristic life can be converted into L_B life as follows:

$$R(t) = e^{-\left(\frac{L_{BX}}{\eta}\right)^{\beta}} = 1 - x. \tag{7.25}$$

7.6 Parametric Accelerated Life Testing

After logarithmic transformation, Eq. (7.25) can be expressed as

$$L_{BX}^{\beta} = \left(\ln\frac{1}{1-x}\right) \cdot \eta^{\beta}. \tag{7.26}$$

If the estimated characteristic life of p value α, η_α, in Eq. (7.24), is substituted into Eq. (7.26), we obtain the B_X life equation:

$$L_{BX}^{\beta} = \frac{2}{\chi_\alpha^2(2r+2)} \cdot \left(\ln\frac{1}{1-x}\right) \cdot \sum_{i=1}^{n} t_i^{\beta}. \tag{7.27}$$

If the sample size is large enough, the planned testing time will proceed as

$$\sum_{i=1}^{n} t_i^{\beta} \cong n \cdot h^{\beta}. \tag{7.28}$$

The estimated lifetime (L_{BX}) in test should be longer than the targeted lifetime (L_{BX}^*):

$$L_{BX}^{\beta} \cong \frac{2}{\chi_\alpha^2(2r+2)} \cdot \left(\ln\frac{1}{1-x}\right) \cdot n \cdot h^{\beta} \geq L_{BX}^{*\beta}. \tag{7.29}$$

Then, sample size equation is expressed as follows:

$$n \geq \frac{\chi_\alpha^2(2r+2)}{2} \cdot \frac{1}{\left(\ln\frac{1}{1-x}\right)} \cdot \left(\frac{L_{BX}^*}{h}\right)^{\beta}. \tag{7.30}$$

However, most lifetime testing has insufficient samples. The allowed number of failures would not have as much as that of the sample size.

$$\sum_{i=1}^{n} t_i^{\beta} = \sum_{i=1}^{r} t_i^{\beta} + (n-r)h^{\beta} \geq (n-r)h^{\beta}. \tag{7.31}$$

If Eq. (7.31) is substituted into Eq. (7.27), B_X life equation can be modified as follows:

$$L_{BX}^{\beta} \geq \frac{2}{\chi_\alpha^2(2r+2)} \cdot \left(\ln\frac{1}{1-x}\right) \cdot (n-r)h^{\beta} \geq L_{BX}^{*\beta}. \tag{7.32}$$

Then, sample size equation with the number of failure can also be modified as

$$n \geq \frac{\chi_\alpha^2(2r+2)}{2} \cdot \frac{1}{\left(\ln\frac{1}{1-x}\right)} \cdot \left(\frac{L_{BX}^*}{h}\right)^{\beta} + r. \tag{7.33}$$

From the generalized sample size Eq. (7.33), we can proceed reliability testing (or parametric ALT testing) under any failure conditions ($r \geq 0$). Consequently, it also confirms whether the failure mechanism and the test method are proper.

7.6.3 Derivation of Approximate Sample Size Equation

As shown in Table 7.4, for a 60% confidence level, the first term $\frac{\chi_\alpha^2(2r+2)}{2}$ in Eq. (7.33) can be approximated to $(r + 1)$ [25]. And if the cumulative failure rate, x, is below about 20%, the denominator of the second term $\ln\frac{1}{1-x}$ approximates to x by Taylor expansion.

Then the general sample size Eq. (7.33) can be approximated as follows:

$$n \geq (r+1) \cdot \frac{1}{x} \cdot \left(\frac{L^*_{BX}}{h}\right)^\beta + r. \tag{7.34}$$

If the acceleration factors in Eq. (7.18) are added into the planned testing time h, Eq. (7.34) will be modified as

$$n \geq (r+1) \cdot \frac{1}{x} \cdot \left(\frac{L^*_{BX}}{AF \cdot h_a}\right)^\beta + r. \tag{7.35}$$

The normal operating cycles of the product in its lifetime are calculated under the expected customer usage conditions. If failed number, targeted lifetime, accelerated factor, and cumulative failure rate are determined, the required actual testing cycles under the accelerated conditions can be obtained from Eq. (7.35). ALT equipment in mechanical/civil system will be designed based on the load analysis and the operating mechanism of the product. Using parametric ALT with approximated sample size of an acceleration factor, the failed samples in the design phase can be found. From the required cycles (or Reliability Quantitative (RQ) test specifications), h_a, it determines whether the reliability target is achieved. For example, without considering the acceleration factor, the calculation results of two sample size equations are presented in Table 7.5.

If the estimated failure rate from the reliability testing is not bigger than the targeted failure rate (λ^*), the number of sample size (n) might also be obtained. The

Table 7.4 Characteristics of $\frac{\chi_{0.4}^2(2r+2)}{2}$ at $\alpha = 60\%$ confidence level

r	$1 - \alpha$	$\frac{\chi_{0.4}^2(2r+2)}{2}$	$\frac{\chi_\alpha^2(2r+2)}{2} \approx r+1$	$1 - \alpha$
0	0.4	0.92	1	0.63
1	0.4	2.02	2	0.59
2	0.4	3.11	3	0.58
3	0.4	4.18	4	0.57

7.6 Parametric Accelerated Life Testing

Table 7.5 The calculated sample size with h = 1080 h testing time

β	Failure number	Sample size	
		Equation (7.33) by Minitab	Equation (7.34)
2	0	3	3
2	1	7	7
3	0	1	1
3	1	3	3

estimated failure rate with a common sense level of confidence (λ) can be described as

$$\lambda^* \geq \lambda \cong \frac{r+1}{n \cdot (\text{AF} \cdot h_a)}. \quad (7.36)$$

By solving Eq. (7.36), we can also obtain the sample size

$$n \geq (r+1) \cdot \frac{1}{\lambda^*} \cdot \frac{1}{\text{AF} \cdot h_a}. \quad (7.37)$$

Multiplying the targeted lifetime (L_{BX}^*) into the numerator and denominator of Eq. (7.37), we can yield another sample size equation:

$$n \geq (r+1) \cdot \frac{1}{\lambda^* \cdot L_{BX}^*} \cdot \frac{L_{BX}^*}{\text{AF} \cdot h_a} = (r+1) \cdot \frac{1}{x} \cdot \left(\frac{L_{BX}^*}{\text{AF} \cdot h_a}\right)^1. \quad (7.38)$$

Here, we know that $\lambda^* \cdot L_{BX}^*$ is transformed into the cumulative failure rate x.

We can see two equations for sample size that have a similar form—Eqs. (7.35) and (7.38). It is interesting that the exponent of the third term for two equations is 1 or β, which is greater than one for wear-out failure. Because the sample size equation for the failure rate is included and the allowed failed number r is 0, the sample size equation Eq. (7.35) for the lifetime might be a generalized equation to achieve the reliability target.

If the testing time of an item (h) is more than the targeted lifetime (L_{BX}^*), the reduction factor R is close to 1. The generalized equation for sample size in Eq. (7.35) might be rewritten as follows:

$$n \geq (r+1) \cdot \frac{1}{x}. \quad (7.39)$$

And if the targeted reliability for module is allocated to B_1 10 years, the targeted lifetime (L_{BX}^*) is easily obtained from the calculation by hand. For refrigerator, the number of operating cycles for one day was 5; the worst case was 9. So the targeted lifetime for 10 years might be 32,850 cycles.

And the other type of sample size equation that is derived by Wasserman [26] can be expressed as

$$n = -\frac{\chi_\alpha^2(2r+2)}{2m^\beta \ln R_L} = \frac{\chi_\alpha^2(2r+2)}{2m^\beta \ln R_L^{-1}} = \frac{\chi_\alpha^2(2r+2)}{2m^\beta \ln(1-F_L^{-1})}$$
$$= \frac{\chi_\alpha^2(2r+2)}{2} \cdot \frac{1}{\ln(1-F_L^{-1})} \cdot \left(\frac{L_{BX}}{h}\right)^\beta, \quad (7.40)$$

where $m \cong h/L_{BX}$, $n \gg r$.

When $r = 0$, sample size equation can be obtained as

$$n = \frac{\ln(1-C)}{m^\beta \ln R_L} = \frac{-\ln(1-C)}{-m^\beta \ln R_L} = \frac{\ln(1-C)^{-1}}{m^\beta \ln R_L^{-1}}$$
$$= \frac{\ln \alpha^{-1}}{m^\beta \ln R_L^{-1}} = \frac{\chi_\alpha^2(2)}{2} \cdot \frac{1}{\ln(1-F_L^{-1})} \cdot \left(\frac{L_{BX}}{h}\right)^\beta. \quad (7.41)$$

So we know that Wasserman's sample size equation Eq. (7.41) is similar to Eq. (7.35).

Especially, the ratio between product lifetime and the testing time in Eq. (7.34) can be defined as reduction factor. It can be used to determine if accelerated life testing is proper. That is,

$$R = \left(\frac{L_{BX}^*}{h}\right)^\beta = \left(\frac{L_{BX}^*}{AF \cdot h_a}\right)^\beta. \quad (7.42)$$

To effectively proceed the parametric accelerated life testing, we have to find the severe conditions that will increase the accelerated factor (AF) and the shape factor β. At that time the location and shape of the failed product in both market and parameter ALT results are similar. If the actual testing time h_a is longer than the testing time that is specified in the reliability target, the reduction faction will be less than one. So we can obtain the accelerated conditions that can decrease the testing time and sample size number.

7.7 The Reliability Design of Mechanical System and Its Verification

7.7.1 Introduction

Completing the design of a new product requires two kinds of activities—managerial and technical skills. Managerial skill includes adopting a process improvement approach, such as Capability Maturity Model Integration (CMMI), and controlling quality, which Japan has pursued for over 60 years (starting with

7.7 The Reliability Design of Mechanical System and Its Verification

training by Dr. W. Edwards Deming at the Japanese Union of Scientists and Engineers) [27]. Technical skill involves using a product-specific validation and verification approach.

CMMI has been developed by the Software Engineering Institute at Carnegie Mellon University [28]. It outlines five organizational levels; from lowest to highest, these are initial, managed, defined, quantitatively managed, and optimizing. The purpose of assessing the level of organization is to raise it to the highest level, at which developing engineer and manager including CEO expect to produce perfect products. This is not an exact approach, but a technique to back up and assess the principal manufacturing process.

Quality control is mainly related to manufacturing. Its focus is how to assure that item variations are within the tolerances of already determined specifications. Therefore, quality control methods are dimensionally different than the verification of new product designs, since the product developer should establish the necessary specifications for new products as well as their tolerances. Quality control is not generally an activity in the design area, but a necessary activity in the manufacturing field in Fig. 7.15.

Let us consider product-specific verification as a technical skill. Generally, engineers check numerous design items when developing new products. In the NASA Systems Engineering Handbook [29], there are 16 activities under the heading "Verification Procedures," almost all of which involve testing. The keywords include identification of test configuration, test objective, test criteria, test equipment, and location of test activity. Similarly, verification of software includes test strategy, test plan, test procedure, test scenario sorting deficiencies, and so on. But these are general comments or recommendations that may vary according to the

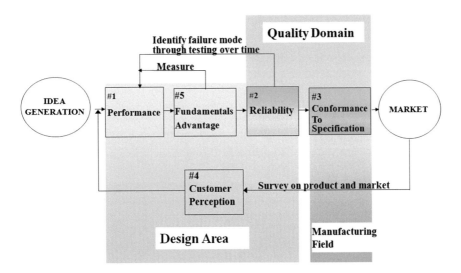

Fig. 7.15 Confusing quality control: (1) R&D, (2) QA (performance/reliability), (3) Manufacture QC

activity and the test article, and therefore are not mandatory. Of course the test is required if applicable specifications exist, but that is not sufficient.

From a verification viewpoint, NASA Handbook addresses tasks used to test products, but does not establish the detailed specification standards as the frame of reference in which these tasks might be carried out. It is not acceptable for verifiers to use their discretion when verifying product performance. Carefully established specifications prevent verification activities to deviate from the determined process. Thus, when failure occurs, it is possible to determine whether the specifications are inappropriate or whether verifiers are incorrectly conforming to the specifications. Sometimes we can also identify omissions in the verification specifications.

Verification specifications should be established over the full range of functions fitted to each product. A thorough use of available technology and related measures to address issues might be applied at an early stage of product development. Why do not developing engineer and manager require verification specifications for each product? The reason lies in the engineers' answers. Product assurance specialists may insist that all related verification activities are included in a "thick document." And they may add that the activities performed are completely reviewed and revised by related specialists. Furthermore, if developing engineer and manager figure out the technical details, he would understand all he wanted to know the design details like differences between quality defects and failures. However, there is a gap between design engineer and manager including CEO. Especially, manager can not understand the complex situations when engineers are in design. This is a kind of trap. If the technical details become specification, the situation will change (Table 4.1).

7.7.2 Reliability Quantitative (RQ) Specifications

Technology concept-related product could be explained with common sense, although some concepts of new technology take time to be understood. Everyone understands new product design concepts if the related staff explains them in everyday communication. Fortunately, the technology concepts related to design verification is not difficult to grasp because we, as consumers, use product. For example, the technology concepts to reduce the noise level of a car engine would be difficult to understand but assessing the performance improvement would be easy since we can hear it. The concepts of design verification-related specification are less complex than that of design itself, and should be easily grasped by developing engineer and manager. If developing engineer and manager has difficulty in understanding the verification document, there are illogical sections not to be explained with common sense when the engineers write it. The necessary logic of "thick document" is clarified for any layman to be understood and controversial as shown in Table 7.6.

Three kinds of simple logic might be addressed in the verification document. First, the information is divided into two activities: verification specification

Table 7.6 Double ambiguity of product quality

Meaning of basic quality	Product Value = Function of Product Quality and Price		
	Reliability		
	Durability	Rate-Reliability	Conformance-Quality
Concept	Product Life	Failure Rate	Conformance to Specifications
Unit	Year	Percent/Year Percent/Hour	Percent ppm
Probability	Weibull distribution	Exponential distribution	Normal distribution
Activities	Design change or establishing specification		Inspection, Screening

establishment and related execution. Second, procedures for how to extract anticipated issues in a new product need to be addressed in the verification specifications, avoiding omissions of necessary specifications and adding some details pertaining to the product for a complete set of specs. Third, verification specifications should be classified into categories according to technological fields in order for related specialists to review their accuracy. Verification specifications might be clearly presented, providing brief summaries to clarify the entire specification concepts.

Additionally, there are two other issues involved in establishing verification specifications. First of all, we think that new products can check the combinations of specifications used for similar products. This is a misunderstanding. Potential problems inherent in new products cannot be identified using old specifications. The new product incorporates innovative structures, new materials, and different softwares for upgrading performance and decreasing cost. These cannot be adequately tested using existing specifications.

Using the previous specifications, new failure mechanisms are not easy to be identified for products that have the design modifications. In addition to updating the specifications, we should also consider what new testing might be effective. For example, is it possible to apply the test specifications for the Boeing 777 fuselage made of aluminum alloys to the Boeing 787 Dreamliner fuselage, which incorporates new materials, like CFRP? Obviously, we know that the previous test specs would be improper.

The other issue is that reliability quantitative specifications that can use the parametric ALT as one of the methodologies mentioned in previous sections include estimating item lifetime. Reliability disasters caused by the design missing during customer use could tarnish the company's reputation. But most people consider this task beyond the scope of possibility. Generally obtaining quantitative results in reliability analysis is very difficult. Reliability specialist Patrick O'Connor wrote in Practical Reliability Engineering that there are basically three kinds of situations—small, moderately large, and very large numbers of factors and interactions [30]. A small number of factors can be predicted with the physical model. A large number can be predicted with statistical models. Predictive power diminishes, however, in the case of a moderately large number of factors pertaining to reliability.

Reliability prediction is a necessary task to be undertaken. Let us look at a product in the standpoint of reliability problems. We know that there are a few sites in product that are weaker than other sites. Reliability specialists can presume the location of the weakest site and/or its failure mechanisms, though they do not know whether the failure will actually happen in the targeted lifetime, or how high the failure rate would be. So if we extract one or two failure sites in the product, mostly in a given module or unit, and classify their failure mechanisms into two categories of reliability—lifetime L_B and failure rate λ within lifetime—the factors related to reliability estimation are decreased, and the cases pertaining to moderately large factors become small-factor-number cases. Thus we can make quantitative estimations about reliability issues—mainly lifetime under normal conditions. This is the simple explanation to understand developing engineer and manager.

Let us describe in commonsense terms the basic concepts of the required statistics and methods pertaining to establish the quantitative lifetime specification, which developing engineer and manager can easily understand the B_X Life as reliability quantitative specifications in Fig. 7.16. For instance, take automobiles. Assume that we test 100 cars in Germany for 10 years and find no trouble (10 years, 160,000 km). We can conclude that the car's failure rate is below 1% per 10 years, which is called "B_1 life 10 years." When we conclude that the car's failure rate is below 1% per 10

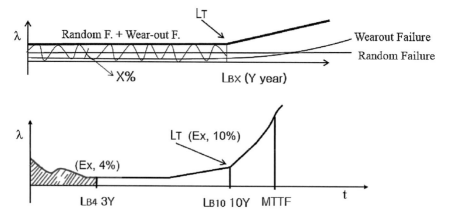

Fig. 7.16 Reliability index; B_X life (L_{Bx})

years, its confidence level reaches around 60%, called the commonsense level of confidence. Of course, we cannot test products for 10 years before market release. So we make the accelerated vehicle testing by imposing heavy loads and high temperature until we reach an acceleration factor of 10. This will reduce the test period by one-tenth, or 1 year. Thus we test 100 items for 1 year (16,000 km), or 1 week without stoppage (7 days × 24 h × 100 km/h = 16,800 km). The next step is to reduce the sample size.

Then, if you increase the testing time, the items would achieve a sufficiently degraded state and many would fail after the test; therefore, we can greatly reduce the sample size, because one or two failed samples would yield enough data to identify the problem area and make corrective action plans. Increasing the test time by four times, or to 1 month, reduces the necessary sample size by the square of the inverse of the test-time multiplier, to one-sixteenth (square of one quarter), or six engines. The final test specification, then, is that six engines should be tested for 1 month under elevated load and temperature conditions with the criterion that no failure is found. This concept, called Parametric Accelerated Life Testing, is the key to reliability quantitative specifications.

We also cannot guarantee the behavior of a product over 10 years under the extreme environments. These test conditions would be appropriate to mechanical/civil system (or components), like power engines, but the test conditions are not fit to assess the degradation of paint on the automobile body. In addition to testing the new engine, we should devise quantitative test methods for other components—new electrical components (including batteries), electronic control units, lighting systems, or coating materials. According to the identified failure mechanisms, testing must be conducted by subassembly to heighten acceleration.

Without such quantitative lifetime testing, we can not identify all the failures influencing the product's lifetime because there would be unanticipated failures. For example, at prototype testing, the lifetime of a tub in a washing machine was lengthened at first parametric ALT by the missing structural design changes—a corner radius increase, rib insertion, and so on. The final parametric ALT, however, showed a weakening of strength in the plastic due to a chemical reaction; the problem was solved by changing the release agent of the injection molding process—something no one could have been predicted as the solution. Note that this method reveals exact failure modes, including totally unexpected ones, that other methods, like FMEA (failure mode and effects analysis), cannot identify.

For the CFRP of the 787 Dreamliner, the failure mechanism is a kind of delamination, which can be found in pressure/humidity/temperature cycling and ultraviolet irradiation testing. If the acceleration factor for testing this is calculated using an adequate life–stress model (time-to-failure model) and the sample size is determined according to the B_X life target, then quantitative results can be derived. Note that we should check the possibility of failure due to various overstresses, such as bird strikes, with sufficiently degraded samples. For the electrical systems in the 787 engineers should incorporate the same components used in other commercial airplanes (and different combinations of them), assessing the possibilities of overstress failures under reliability marginal stresses, since they can assure lifetime reliability. But for new components like the lithium-ion battery, the failure mechanics as well as the stresses produced in the aircraft environment have been changed. Thus, we cannot presume what kind of failure mechanisms would occur due to chemical reaction until the projected lifetime reaches. Generally, chemical failure mechanisms are delicate and thus difficult to identify and reproduce, which means that the acceleration conditions and related factors can be hard to determine. Thus they should test until lifetime under almost normal conditions, and the behavior of sufficiently degraded components should be checked under the rated stresses and overstresses. The media have reported various accidents or disasters due to unanticipated failure mechanisms in chemical items, such as the fires occurring in the Sony lithium-ion battery in notebook computers in 2005, Firestone tires causing Ford Explorer rollovers in the 1970s, wire bundles incorporating silver-plated copper wire leading to fire in the Apollo 1 cabin in 1967, and so on.

7.7.3 Conceptual Framework of Specifications for Quality Assurance

Returning to the subject of establishing verification specifications, there are plenty of specifications that have few explanations technically. It is difficult to find articles that explain to establish specifications; there have been a few research studies about it. So let us consider how to anticipate issues in a new product and to configure a series of verification specifications responding to them, and how to develop specifications that will identify these issues accurately.

Here is one such methodology. First, select an existing product to be compared with the new one. All its relevant specifications are listed except the unnecessary specifications. Second, because inevitably the similar but older product has the intractable problems to be listed, we must devise new specifications to address these issues. Ongoing problems indicate that any countermeasures have not resolved the real cause because of the inadequate analysis. Nonetheless, the original design idea may be faulty. The existing product would be solved using precise problem analysis, and the new product would be handled by identifying and fixing the problem before releasing the next model. To correct the existing problems in similar products, it is important to add the verification specifications. Sometimes the potential problems of the subassemblies manufactured by a new supplier also might be considered.

Third, the newly designed portions—those that differ from the current comparable product—should be listed and the potential issues related to them should be predicted. Verification specifications need to be devised to address these problems. Especially note that all items incorporating new chemical materials should be tested to item lifetime under new quantitative specifications because a new kind of

7.7 The Reliability Design of Mechanical System and Its Verification

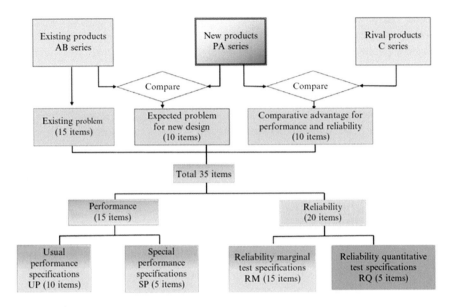

Fig. 7.17 Complete testing sets of verification specifications for quality assurance

wear-out failure could occur near the item lifetime. Moreover, it is very difficult to computer simulate and clarify diverse chemical reactions over an item's life cycle.

Finally, the new product will also have performance fundamentals unique to it, which sometimes provide a competitive edge over competitors. Such comparative advantages in performance fundamentals might be checked with the newly established specifications.

A complete set of verification specifications in Fig. 7.17 might be summarized as four types of data: (1) all the verification specifications for the comparable product(s); (2) specifications to fix existing problems in the comparable product; (3) specifications that deal with the potential issues in the newly designed portions; and (4) specifications checking newly incorporated performance features. The specifications responding to the latter three categories are all established anew. The purpose of sectioning potential issues in a new product is to check whether necessary issues have been omitted.

All specifications enumerated according to this model would be classified, initially, into two groups: performance specifications and reliability specifications. If an issue to be identified is related to material rupture or degradation over time, it is a reliability issue; if not, it is a performance issue. The specifications are further divided into four categories: Usual Performance specifications (UP), Special Performance specifications (SP), Reliability Marginal test specifications (RM), and Reliability Quantitative test specifications (RQ).

UPs check the expected performance by the usual operator or consumer. SPs check performance under extraordinary environments, such as tropical heat or elevated electromagnetic fields. RMs are used for identifying physical changes under severe or peculiar conditions, including unusual usage environments like electrostatic overstress or lightning surges. Finally, RQs are for reviewing the product state under normal conditions and for estimating the product lifetime, the B_X life (lifetime of the cumulative failure rate X %), and the annual failure rate within lifetime. Note that the lifetime index MTTF refers to the time point at which about 60% of the production lot fails, which is an unacceptable rate.

Parametric accelerated life testing mentioned in the previous sections uses the sample size equation with acceleration factor. It also is a process that helps designers find the optimal design parameters, which can help them better estimate expected lifetime L_B, failure rate of module λ, and determine the overall product reliability. Reliability quantitative (RQ) test specifications are used to estimate the required lifetime (or cycle) if reliability target of product—the cumulative failure rate X % and lifetime is given. Parametric accelerated life testing (ALT) might be related to the RQ test specifications. And the examples of parametric ALT will be discussed with Chap. 8.

7.8 Testing Equipment for Quality and Reliability

7.8.1 Introduction

In today's competitive market, more companies are looking to application-specific automatic testing equipment versus functional testing methods. That is because the traditional testing process did not apply for complex systems such as aero and automotive engines. High product performance and reliability are a basic requirement and sometimes the only difference between products of various manufacturers. Test equipment verifies the performance and reliability of mechanical, electrical, hydraulic, and pneumatic products. These include tool testers, hi-pot testers, power cord and power supply cord testers, automatic test equipments for a variety of purpose, and leakage current testers.

Product quality is a critical aspect for companies who are struggling to retain customers in these days of fast eroding brand loyalty. Testing equipment companies designs and builds production test equipment. They specialize in R&D test equipment, authentication test equipment, and quality control test equipment for mass production. Test equipment is categorized as overall performance test, durability (life) test, accelerated test, safety test and environmental test, etc. (see Fig. 7.18).

Test equipment has multi-disciplinary systems that are incorporating machine design, material science, industrial engineering, statistics, electrical & electronics, and computing system. As product development requires substantially high level of

7.8 Testing Equipment for Quality and Reliability

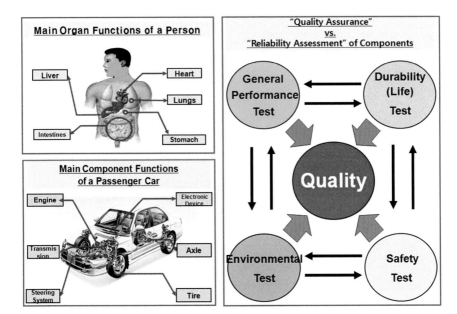

Fig. 7.18 Reliability assessment concept for developing testing equipment

performance and reliability in the limited developing time of product, equipment for testing the performance and reliability of product is growing at a significant scale.

In the stage of the detail design, testing equipment companies are to develop the test equipment that is applicable to be gratified at the specifications of end users. They provide the latest state-of-the-art test equipment to rental centers, electrical service facilities, manufacturers, and OEMs. In over half century they have learned what end users want. Their experiences are reflected in a number of important concepts in hardware and software designs of test equipment. When a request for a test equipment development is received, testing equipment companies promises to provide customized support by analyzing the necessary requirements for installation and trial runs of the developed equipment.

The Quality, Safety, and Life of product could be increased by reliability testing. Product reliability testing is a specialized field that requires deep understanding of the product and a state-of-the-art infrastructure to deliver the goods. Product reliability testing equipment is to help companies to test reliability of their products. Thus, reliability testing equipment should have a user requirement—purpose, required power, testing items, control precision, data processing speed, automation level, software processing ability, maintainability, equipment maintenance cost, spare part, and necessary budget (see Fig. 7.19).

Testing equipment insures the reliability, safety, and performance of products they manufacture, use, service, or rent. As product technology advances, testing equipment are required to (1) make products reliable, (2) meet the international standards, and (3) offer the Product/Parts Reliability, Failure Analysis, Test

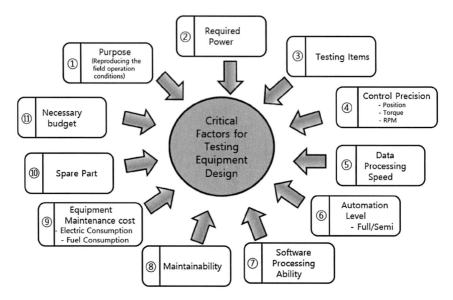

Fig. 7.19 Considerations for developing testing equipment

Structure (Design, Verification and Test), Technology Qualification Support for product, and environmental Measurement Services (humidity, temperature, etc.).

Consequently, testing equipment type for product R&D Development can be classified as follows: (1) Testing Equipment for General Performance, (2) Testing Equipment for Durability (Life), (3) Testing Equipment for Accelerated Testing, (4) Testing Equipment for (Combined) Environment, (5) Testing Equipment for Quality Control of Mass Production, and (6) Testing Equipment for Maintenance and Repair. As shown in Fig. 7.20, there are a variety of types of testing equipment and their company in global that cannot be quantified.

Today testing equipment companies also look to custom-designed and manufactured automatic testing equipment that can functionally test new units that employ advanced technologies. By going beyond simple parametric testing that limits the use of commercial off the shelf testers, specialty-built functional automatic testing equipment helps guarantee high intrinsic availability and long-lived performance "to spec" in the field, thereby facilitating the acceptance and success of new technologies in the marketplace.

They often specialize in selling test equipment and offering a variety of services to survive the marketplace. They buy, sell, and lease all kinds of new, refurbished, and used equipments. They also buy networking equipment, used test equipment, and used measurement equipment from leading manufacturers. Whether end users are any reseller, they offer a cost-effective solution that will save time and money.

7.8 Testing Equipment for Quality and Reliability

Fig. 7.20 Type of quality testing equipment

7.8.2 Procedure of Testing Equipment Development (Example: Solenoid Valve Tester)

Development procedure of testing equipment can be briefly summarized in Fig. 7.21. For example, the test equipment of solenoid valve tester in nuclear power plant will be suggested. The testing equipment would test the intended functionality of product and its reliability.

Step 1. **Characteristic Study of Product to be tested**

A solenoid valve is operated by an electric current through a solenoid. For more than 440 nuclear power plants in the world, a solenoid valve has equipped nearly every plant. As seen in Fig. 7.22, nuclear-qualified and critical solenoid valves have the following applications:

- Emergency core cooling systems
- Emergency generator systems
- Steam generator feed-water systems
- Containment sampling systems
- Auxiliary feed-water systems
- Liquid radiation waste systems
- Turbine bypass systems

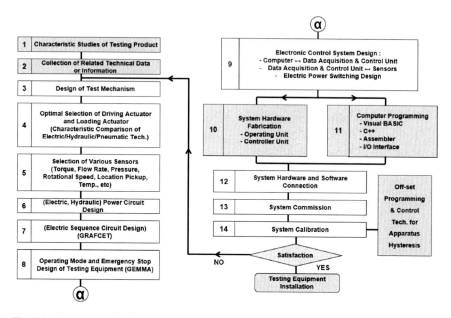

Fig. 7.21 Procedure of testing equipment development

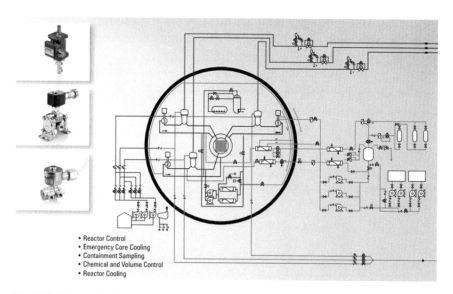

Fig. 7.22 Usage of solenoid valve in nuclear power plant

7.8 Testing Equipment for Quality and Reliability

Nuclear-qualified solenoid valves are indispensable parts of any nuclear plant safety application. Each has passed the most rigorous testing for nuclear equipment and environmental qualification (EQ). These solenoid valves are produced with a high degree of designed-in quality and proven performance.

Additionally, solenoid valves offer desirable product advantages such as diodes that provide simple surge protection for control, quick-disconnect connectors for increased safety and reduced maintenance, and radiation-resistant elastomers for long life.

In the 1950s solenoid valves were onboard the first nuclear-powered submarine, the USS Nautilus. Later, solenoid valves protected the earliest commercial nuclear power plants. In 1978 solenoid valves were among the first and only to be nuclear qualified to IEEE and RCC-E specifications. Solenoid valves from specialized nuclear line are specifically designed for environments with high radiation, temperature, and seismic requirements.

Consequentially, test equipment for a solenoid valve should satisfy the following specifications:

- ISO 6358: 1989 (E) Pneumatic fluid power—Components using compressible fluids—Determination of flow rate characteristics
- IEEE-323: 2003—Standard for Qualifying Class 1E Equipment for Nuclear Power Generating Stations

Step 2. **Collect-Related Technical Information/Data**

As seen in Figs. 7.23 and 7.24, nuclear solenoid valves meet the rigorous demands and high expectations of the nuclear industry. They have applications for nuclear-qualified 2-, 3-, and 4-way solenoid valves. Especially, nuclear 2-way valves are qualified for mild environmental applications as defined in IEEE-323-2003. The qualification program consisted of a series of four sequential aging simulation phases (thermal, wear, radiation, and vibration).

- IEEE-323: 2003—Standard for Qualifying Class 1E Equipment for Nuclear Power Generating Stations.

Qualification consists of subjecting solenoid valve to the following tests as required by the previously noted IEEE-323 specifications.

A. Thermal aging
B. Wear aging
C. Pressurization aging
D. Radiation aging
E. Vibration aging
F. Seismic event simulation
G. Radiation event simulation
H. LOCA/MSLB/HELB environmental simulation

Fig. 7.23 Type of solenoid valve for nuclear power plant

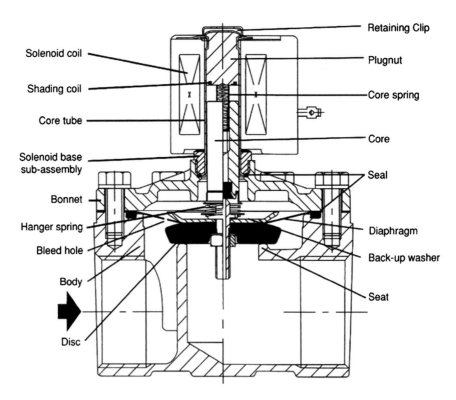

Fig. 7.24 Structure of solenoid valve for nuclear power plant

7.8 Testing Equipment for Quality and Reliability

All solenoid valves are designed with the following special features:

1. Type N Construction (NS Series), Class H (NP Series) coil insulation.
2. Elastomers (gaskets, O-rings, disks): all materials designed to meet high radiation and high temperature degradation effects.
3. Specially designed solenoid enclosures to withstand Loss-of-Coolant-Accident (LOCA) environment.
4. Designed to meet seismic loading.

2-way Nuclear Power (NP) solenoid valves are widely used for pilot control of diaphragm and cylinder actuated valves (and other applications) used in nuclear power plants. Selection of the proper valve for a specific application is of paramount importance. This engineering information section describes principles of operation, types of solenoid valves, and types of solenoid enclosures, and materials to assist you in the proper selection of a valve.

A nuclear solenoid valve is a combination of two basic functional units: (1) a solenoid, consisting of a coil and a magnetic plunger (or core); and (2) a valve body containing an orifice in which a disk is positioned to stop or allow flow. The valve is opened or closed by movement of the magnetic plunger (or core), which is drawn into the solenoid when the coil is energized. Solenoid valves feature a packless construction. The solenoid is mounted directly on the valve and the core assembly is enclosed in a sealed tube inside the solenoid coil. This construction provides a compact, leak-tight assembly, without the need of a stuffing box or sliding stem seal.

Direct acting solenoid valves operate from zero kPa (no minimum pressure is required for the valve to operate) to the individual valve's maximum rated pressure. Because of the wide range of sizes, construction materials, and pressures, direct-acting qualified valves in brass or stainless steel are found to the many applications found in nuclear power plants. Two 2-way direct acting types are available as follows: normally Closed: closed when de-energized and open when energized; and normally Open: open when de-energized and closed when energized.

Step 3. Design of Test Mechanism

Method to determinate flow rate characteristic of the solenoid valve is based on increasing upstream pressure while the pressurized air goes through a mass flow sensor. The method of standard ISO 6358 is explained in two equations which describe the flow rate through the orifices:

$$q_m = C p_1 \rho_0 \sqrt{\frac{T_0}{T_1}} \quad \text{for} \quad \frac{p_2}{p_1} \leq b \text{ choked flow} \quad (7.43)$$

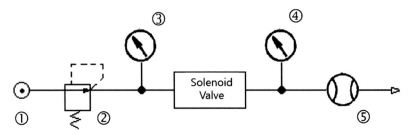

(a) Schematic of measuring the flow characteristics of solenoid valve (1 – pressurized air source, 2 – pressure regulator, 3/4 – pressure sensor, 5 – flow meter)

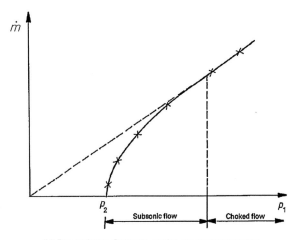

(b) Dependence flow rate on the upstream pressure.

Fig. 7.25 Measuring flow rate characteristic by increasing upstream pressure

$$q_m = Cp_1\rho_0\sqrt{\frac{T_0}{T_1}}\sqrt{1-\left(\frac{\frac{p_2}{p_1}-b}{1-b}\right)^2} \quad \text{for} \quad \frac{p_2}{p_1} \succ b \text{ subsonic flow.} \quad (7.44)$$

Custom equipment was manufactured for measuring the flow rate characteristics based on "increasing the upstream pressure." Part of the valve holding the valve nozzle was replaced with the special equipment. With this setup it was possible to set the nozzle to a required fixed position. Without modifications this could not be done.

Figure 7.25a shows the measuring station which was used for measurement and Fig. 7.25b shows the curvature of dependence flow rate on the upstream pressure. This method has been applied only on solenoid valve because preparing special equipment for holding the nozzle in a constant position is very expensive and work intensive. Step 4–13.

7.8 Testing Equipment for Quality and Reliability

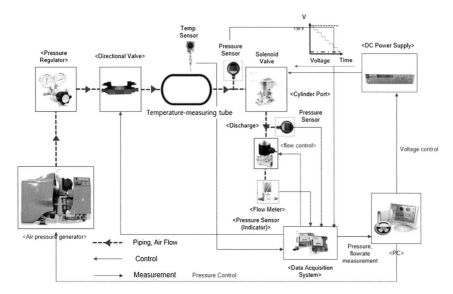

Fig. 7.26 Schematic control diagram of measuring flow rate characteristics of solenoid valve

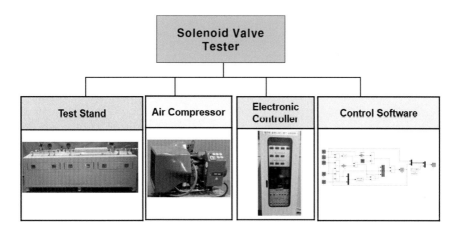

Fig. 7.27 Solenoid valve tester

Step 4–13. **Making the solenoid testing equipment**

These steps consist of optimal setting of driving actuator and loading actuator, selection of various sensors, design power (electric/hydraulic) circuit, design sequence circuit, design operating mode, design automatic stop mode, electronic control system design, computer ↔ DAQ & control unit, DAQ & control unit ↔ sensor, electric power switching circuit design, fabrication of system hardware,

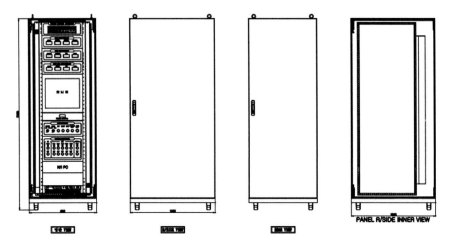

Fig. 7.28 Appearance of electronic controller

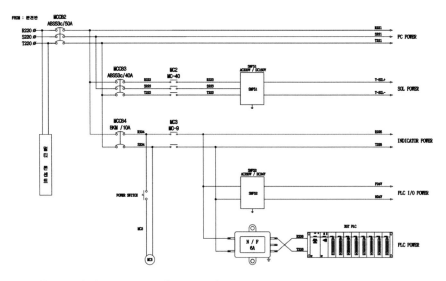

Fig. 7.29 Electrical power connection (example)

fabrication of operating unit, fabrication of controller, computer programming by LabVIEW and MATLAB, and combine system hardware and software (see Figs. 7.26, 7.27 and 7.28).

For creating a simulation of pneumatic fluid power—determination of flow rate characteristics of solenoid valve using compressible fluids—MATLAB® Simulink may be used. It makes possible to compare a lot of measured data with mathematical models, which was a great contribution to the work. Measurements made with the two valves were compared to theoretical values (Fig. 7.29).

7.8 Testing Equipment for Quality and Reliability

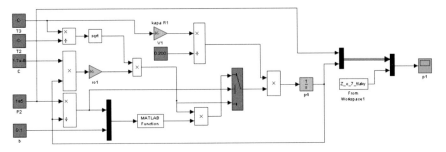

Fig. 7.30 Simulation schematic of a tank charge in MATLAB® Simulink

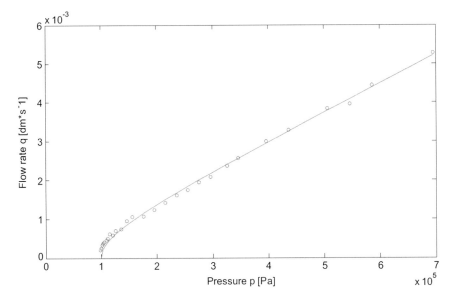

Fig. 7.31 Dependence of flow rate to the upstream pressure—data from the ISO 6358 method measurement

Figure 7.30 shows two Simulink models created to determine sonic conductance C and the critical pressure ratio b from the measured data. The flow rate characteristic was measured by increasing upstream pressure as depicted in figure.

Step 14–15. **System simulation, test run, and calibration**

The result is shown in Fig. 7.31. Measured data are represented as circles. Line curvature represents theoretical approximation obtained as a numerical solution searching sonic conductance C and the critical pressure ratio b. The results are $C = 6.329 \times 10^{-9}$ m³/(s Pa), $b = -0.1527$, which can be obtained from MATLAB® and indicates the solenoid valve characteristics. And this method might be applied only on solenoid valve because preparation of the special equipment for holding the nozzle in constant position is very expensive and work intensive.

Values of critical pressure ratio b for each final pressure in the tank were determined based on knowledge that the linear part of the curvature of the charge is describe by Eq. (7.1) and the behaviors of flow rate in the second part of the curvature are described by Eq. (7.2). Of note is that the critical pressure ratio b expresses the divide of the downstream and the upstream pressure which flow becomes choked.

For determination of the critical pressure ratio a derivative of smoothed measured data was made, and the point where the derivative exchanges determines the mentioned critical pressure ratio. The values of sonic conductance C were determined from the equation which describes the flow rate through the orifices for subsonic flow.

References

1. 1990 IEEE Standard Glossary of Software Engineering Terminology (1990) IEEE STD 610.12-1990. Standards Coordinating Committee of the Computer Society of IEEE, New York
2. Kreyszig E (2006) Advanced engineering mathematics, 9th edn. Wiley, Hoboken, NJ, p 683
3. Taguchi G (1978) Off-line and on-line quality control systems. In: Proceedings of the international conference on quality control, Tokyo, Japan
4. Taguchi G, Shih-Chung T (1992) Introduction to quality engineering: bringing quality engineering upstream. ASME, New York
5. Ashley S (1992) Applying Taguchi's quality engineering to technology development. Mechanical Engineering
6. Wilkins J (2000) Putting Taguchi methods to work to solve design flaws. Qual Progress 33(5):55–59
7. Phadke MS (1989) Quality engineering using robust design. Prentice Hall, Englewood Cliffs, NJ
8. Byrne D, Taguchi S (1987) Taguchi approach to parameter design. Qual Progress 20(12): 19–26
9. Woo S, Pecht M (2008) Failure analysis and redesign of a helix upper dispenser. Eng Fail Anal 15(4):642–653
10. Woo S, O'Neal D, Pecht M (2009) Improving the reliability of a water dispenser lever in a refrigerator subjected to repetitive stresses. Eng Fail Anal 16(5):1597–1606
11. Woo S, O'Neal D, Pecht M (2009) Design of a hinge kit system in a Kimchi refrigerator receiving repetitive stresses. Eng Failure Anal 16(5):1655–1665
12. Woo S, Ryu D, Pecht M (2009) Design evaluation of a French refrigerator drawer system subjected to repeated food storage loads. Eng Fail Anal 16(7):2224–2234
13. Woo S, O'Neal D, Pecht M (2010) Failure analysis and redesign of the evaporator tubing in a kimchi refrigerator. Eng Fail Anal 17(2):369–379
14. Woo S, O'Neal D, Pecht M (2010) Reliability design of a reciprocating compressor suction reed valve in a common refrigerator subjected to repetitive pressure loads. Eng Fail Anal 7(4):979–991
15. Woo S, Pecht M, O'Neal D (2009) Reliability design and case study of a refrigerator compressor subjected to repetitive loads. Int J Refrig 32(3):478–486

16. Woo S, O'Neal D, Pecht M (2011) Reliability design of residential sized refrigerators subjected to repetitive random vibration loads during rail transport. Eng Fail Anal 18(5):1322–1332
17. Woo S, Park J, Pecht M (2011) Reliability design and case study of refrigerator parts subjected to repetitive loads under consumer usage conditions. Eng Fail Anal 18(7):1818–1830
18. Woo S, Park J, Yoon J, Jeon H (2012) The reliability design and its direct effect on the energy efficiency. In: Energy efficiency—the innovative ways for smart energy, the future towards modern utilities, Chap 11. InTech
19. Woo S (2015) The reliability design of mechanical system and its parametric ALT. In: Handbook of materials failure analysis with case studies from the chemicals, Chap 11. Concrete and Power Industries. Elsevier, Amsterdam, pp 259–276
20. Woo S, O'Neal D (2016) Improving the reliability of a domestic refrigerator compressor subjected to repetitive loading. Engineering 8:99–115
21. Woo S, O'Neal D (2016) Reliability design of the hinge kit system subjected to repetitive loading in a commercial refrigerator. Challenge J Struct Mech 2(2):75–84
22. McPherson J (1989) Accelerated testing. In: Packaging, electronic materials handbook, vol 1. ASM International, pp 887–894
23. Karnopp DC, Margolis DL, Rosenberg RC (2012) System dynamics: modeling, simulation, and control of mechatronic systems, 5th edn. Wiley, New York
24. Ajiki T, Sugimoto M, Higuchi H (1979) A new cyclic biased THB power dissipating ICs. In: Proceedings of the 17th international reliability physics symposium, San Diego, CA
25. Ryu D, Chang S (2005) Novel concept for reliability technology. Microelectron Reliab 45(3):611–622
26. Wasserman G (2003) Reliability verification, testing, and analysis in engineering design. Marcel Dekker, p 228
27. Deming WE (1950) Elementary principles of the statistical control of quality. JUSE, Japan
28. CMMI Product Team (2002) Capability Maturity Model Integration (CMMI) Version 1.1, continuous representation. Report CMU/SEI-2002-TR-011. Software Engineering Institute, Pittsburgh PA
29. NASA (2007) System engineering handbook. NASA Headquarters (NASA/SP-2007-6105 Rev 1), Washington, p 92
30. O'Connor P (2002) Practical reliability engineering. Wiley, New York, p 159

Chapter 8
Parametric ALT and Its Case Studies

Abstract In this chapter, parametric accelerating life testing (ALT) and its case studies that can be applicable to a variety of mechanical product will be discussed. Because parametric ALT described in Chap. 7 is suggested as the methodology of the reliability-embedded product developing process, it is important for engineer to figure out how to apply this method to the mechanical system. Here examples of mechanical systems are applicable to airplane, automobiles, construction equipment, washing machines, and vacuum cleaners. To meet the targeted reliability of mechanical product (or module), parametric ALT can identify the missing controllable design parameters. After a tailored series of accelerated life tests, new product will satisfy the reliability target because there is no missing design parameter.

Keywords Parametric ALT · Case studies · Mechanical engineering system

8.1 Failure Analysis and Redesign of Ice Maker

The basic function of a refrigerator is to store fresh and/or frozen foods. Today refrigerators also provide other functions—dispensing ice and water. As the number of refrigerator parts and their functions increase, market pressure for product cost reduction leads to the use of cheaper parts. At that time refrigerator functions are consistently reliable during customer usage. The refrigerator can be designed for reliability by determining proper parameters and their levels.

However, minor design parameters may be neglected in the design review, resulting in product failure in use. Products with minor design flaws may result in recalls and loss of brand name value. Furthermore, product liability law requires manufacturers to design products more safely in the European Union and the United States. Preventing such outcomes is a major objective of the product development process—design, production, shipping, and field testing.

Conventional methods, such as product inspection, rarely identify the reliability problems occurring in market use. Instead, optimally designing for reliability requires the extensive testing at each development step. As a result, the cost of quality assurance and appraisal can increase significantly. As a solution, most global companies focus on accelerating life testing (ALT). ALT can help shorten the product development cycles and identify diverse design flaws. ALT should be performed with sufficient samples and testing time with equipment designed to match expected product loads.

Figure 8.1 shows the SBS refrigerator with ice dispenser and the mechanical parts of the ice bucket assembly. The assembly consists of the bucket case, helix support, helix dispenser clamp, blade dispenser, helix upper dispenser, and blade, as shown in Fig. 8.1b. The helix upper dispenser in the ice bucket of refrigerators with ice dispenser systems has been fracturing in field, causing loss of the dispensing function (see Figs. 8.2 and 8.3). Thus reproducing the failure mode to assess how to prevent the fracture of the helix upper dispenser was critical. The data on failed products in the marketplace are important to understand the use environment of customer of the product and helping to pinpoint root causes.

To investigate the reliability of a helix upper dispenser, using robust design schematic, Bond graph, and state equations analyzed "uncontrollable" mechanical load conditions of an ice bucket assembly New ALT methodologies was proposed for robust designs.

Field data indicated that the damaged products may have two structural design flaws: (1) a 2 mm gap between the blade dispenser and the helix upper dispenser, and (2) a weld line around the impact area of the helix upper dispenser. Due to the gap, the rotating blade dispenser impacts the fixed helix upper dispenser. Because of the weld line, a crack may occur. The temperature of the product was below $-20\ ^\circ C$.

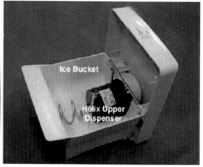

(a) SBS refrigerator   (b) Mechanical parts of the ice bucket assembly

Fig. 8.1 SBS refrigerator and ice bucket assembly

8.1 Failure Analysis and Redesign of Ice Maker

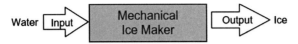

Fig. 8.2 Robust design schematic of ice maker

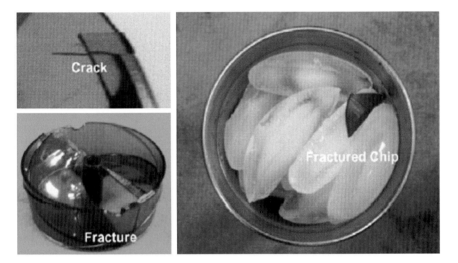

Fig. 8.3 A damaged product after use

As seen in Fig. 8.4, the bond graph can be conventional in state space representation to group terms by state variables. The modeling of ice bucket assembly can be expressed as

$$\begin{bmatrix} di_a/dt \\ d\omega/dt \end{bmatrix} = \begin{bmatrix} -R_a/L_a & 0 \\ mk_a & -B/J \end{bmatrix} \begin{bmatrix} i_a \\ \omega \end{bmatrix} + \begin{bmatrix} 1/L_a \\ 0 \end{bmatrix} e_a + \begin{bmatrix} 1 \\ -1/J \end{bmatrix} T_D \quad (8.1)$$

The mechanical stress (or life) of the ice bucket assembly depends on the disturbance load T_{Pulse} in Eq. (8.1). The accelerated life testing applies the stress between low and high to the breakdown stress. The life–stress model (LS model) can be modified as

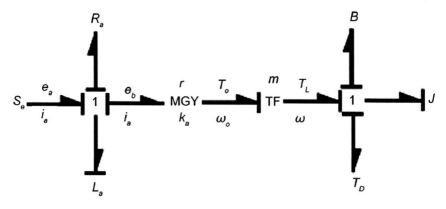

Fig. 8.4 Bond graph of ice bucket assembly

$$\text{TF} = A(S)^{-n} = A(T_D)^{-\lambda} \tag{8.2}$$

The acceleration factor (AF) can be derived as

$$\text{AF} = \left(\frac{S_1}{S_0}\right)^n = \left(\frac{T_1}{T_0}\right)^\lambda \tag{8.3}$$

The ice dispenser of customer is used an average of approximately 3–18 times per day. Under maximum use for 10 years, the dispenser incurs about 65,700 usage cycles. Data from the motor company specifies that normal torque is 0.69 kN cm and maximum torque is 1.47 kN cm. Assuming the quotient $n = 2$, the acceleration factor is approximately 5 in Eq. (8.3).

The test cycles and test sample numbers calculated in Eq. (7.36) were 42,000 cycles and 10 pieces, respectively. The parametric ALT was designed to ensure a B1 of 10 years life with about a 60% level of confidence that it would fail less than once during 42,000 cycles.

Figure 8.5 shows the ALT equipment for the reproduction of the failed structural parts in the field and the duty cycles for the disturbance load T_D. Figure 8.6 shows the failed product in the field and a sample after accelerated life testing. In the photo, the shape and location of the broken pieces in the failed market product are identical to those in the ALT results. Figure 8.7 represents the graphical analysis of the ALT results and field data on a Weibull plot. For the shape parameter, the estimated value in the first ALT is 2.0. However, the final value obtained on the Weibull plot was 4.8. As the ratio of characteristics life, η_1/η_2, gives the acceleration factor, AF is approximately 2.2 on the Weibull plot.

These methodologies are valid to reproduce the fielded failures because (1) the location and shape of the fractures in both market and ALT results are extremely similar; and (2) on the Weibull, the shape parameters of the ALT results, $\beta 1$ and market data, $\beta 2$, are very similar. The reduction factor R also is 0.001 from the

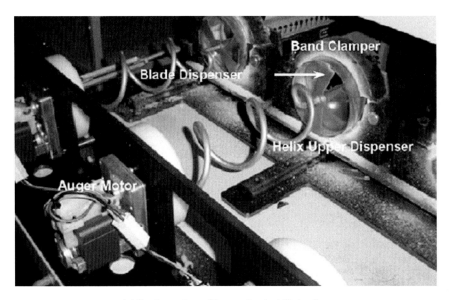

(a) Equipment used in accelerated life testing

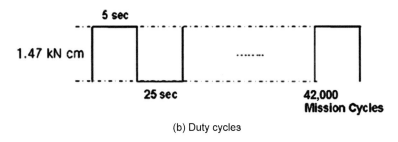

(b) Duty cycles

Fig. 8.5 Equipment used in accelerated life testing and duty cycles of disturbance load T_{Pulse}

experiment data—product lifetime, acceleration factor, actual mission cycles, and shape parameter. Consequently, we know that this parameter ALT is effective to save the testing time and sample size.

The fracturing and cracking of both the fielded products and the ALT results occur in the contact area of the blade dispenser. These structural flaws generate the concentrated mechanical stress when the blade dispenser, made of stainless steel, meets the polycarbonate helix upper dispenser at a right angle. Due to the 2 mm gap between the blade dispenser and helix upper dispenser and the impact (1.47 kN cm) of the blade dispenser, the concentrated stress of the blade dispenser is approximately 36.9 kPa, based on finite element analysis. Under −20 °C, it is particularly fragile due to the weld line near the impact area of the helix upper dispenser (Fig. 8.8).

226　　8 Parametric ALT and Its Case Studies

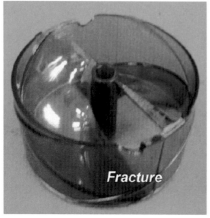
(a) Failed product in field

(b) Failed sample in accelerated life testing

Fig. 8.6 Failed product in field and ALT

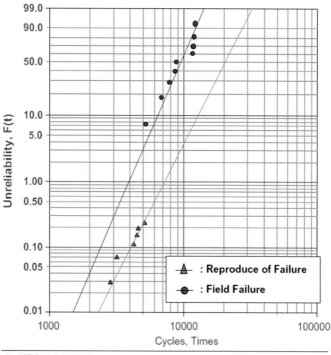

$\beta1=4.7785, \eta1=1.0262E+4$
$\beta2=4.0710, \eta2=2.2215E+4$

Fig. 8.7 Field data and results of ALT on Weibull chart

8.1 Failure Analysis and Redesign of Ice Maker

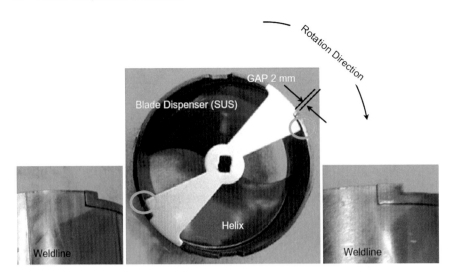

Fig. 8.8 Structure of helix upper dispenser

Table 8.1 show the improved design of the helix upper dispenser based on the ALT results. Failure analysis identified the root cause of the failed product as the 2 mm gap between the blade dispenser and the helix upper dispenser, and the weld line. To improve the reliability of the newly designed helix upper dispenser, a second ALT was implemented with a key controllable design improvement—no gap in the samples. Based on the first ALT, the AF and β values in the second ALT were 2.2 and 4.8. The test cycles and test sample number calculated in Eq. (7.36) were 54,000 cycles and 6 pieces, respectively. For the second ALT, all samples were failed within 54,000 cycles. In the second ALT results the failed test samples were still found in mission test cycles.

For the failed samples, the key controllable design improvement in the third ALT was to add ribs on the side and front of the impact area. These redesigned samples were implemented for the third ALT. The test cycles and test sample calculated in Eq. (7.36) were 54,000 cycles and 6 pieces, respectively. The ALT was designed to ensure a B1 of 10 years life with about a 60% level of confidence that it would fail less than once during 54,000 cycles. In the third ALT results, the samples did not crack and fracture until 75,000 cycles of testing. Consequently, the improved helix upper dispenser will meet the reliability target—B1 10 years.

Table 8.1 shows the results obtained from the third ALT. The B1 life of the redesigned samples was 14 years. When the design of the current product was compared with that of the newly designed one, the B1 life expanded about fourteen times, from 1.4 to 14 years. The design improvements of eliminating the gap and reinforcing the ribs were very effective in enhancing the reliability of the samples (see Fig. 8.9).

Table 8.1 Results of ALT

	1st ALT	2nd ALT	3rd ALT
	Initial design	Second design	Final design
In 54,000 cycles, no crack and fracture of helix	170 cycles: 1/10 (10%) 5200 cycles: 1/10 (20%) 7880 cycles: 2/10 (40%) 8800 cycles: 2/10 (60%) 11,600 cycles: 4/10 (100%)	17,000 cycles: 1/6 (17%) 25,000 cycles: 3/6 (67%) 28,200 cycles: 1/6 (83%) 38,000 cycles: 1/6 (100%)	54,000 cycles: OK Max 75,000 cycles: OK
Helix upper dispenser structure			
Material and specification	PC + SUS ($t = 1.2$) Gap C1: 2 mm $\rightarrow$ 0 mm	Roundness corner of torsional shaft C2: R0.5 mm $\rightarrow$ R2.0 mm	PC + SUS ($t = 1.2$) Gap: 0 mm Added rib on side and front of helix

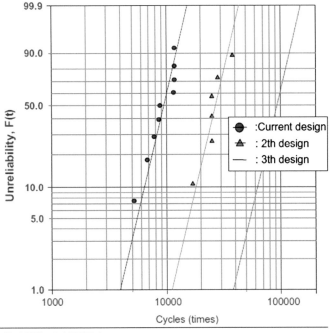

$\beta_1=4.7785, \eta_1=1.0262E+4$
$\beta_2=4.7800, \eta_2=2.9069E+4$
$\beta_3=4.7800, \eta_3=1.0024E+5$

Fig. 8.9 Result of ALT plotted in Weibull chart

8.2 Residential Sized Refrigerators During Transportation

Figure 8.10 shows a typical residential sized French door refrigerator and the mechanical compartment at the bottom rear of the refrigerator. As refrigerators were transported to the final destinations by rail, they were subjected to random vibrations from the train. These vibrations were continually transmitted to the refrigerator (or machine compartment) while train was moving. The connecting tubes in the mechanical compartments of refrigerators were fracturing and the compressor rubber mounts were tearing. Because the tubes were fracturing, refrigerant was leaking out of the tubes, which resulted in the refrigerator losing its ability to either cool or freeze products. Field data indicated that the damaged products might have had design flaws. The design flaws combined with the repetitive random loads could cause failure.

Based on the field data, the rail transportation was expected to move a refrigerator 7200 km from Los Angeles to Boston in 7 days (L_B^*). For its machine compartment (or module), B1 life should be kept for the transported distance (see Fig. 8.11).

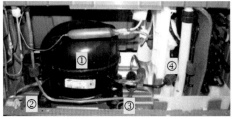

(a) French Door Refrigerator   (b) Machine compartment: (1) compressor, (2) rubber, (3) connecting tubes, and (4) fan and condenser

Fig. 8.10 French door refrigerator and machine compartment (or module)

A random vibration in refrigeration system is motion which is nondeterministic. Refrigerator is subjected to ride on a rough road or rail, wave height on the water. A measurement of the acceleration spectral density is the usual way to specify random vibration. As seen in Fig. 8.12, a refrigerator subjected to base is random vibrations and their power spectral density.

A refrigerator subjected to random vibration during transportation can be modeled using the one-degree-of-freedom vehicle model (see Fig. 8.13). The equivalent model of refrigerator is simplified as:

$$m\ddot{x} + c\dot{x} + kx = ky + c\dot{y} \tag{8.4}$$

The force transmitted to the refrigerator can be expressed as force transmissibility Q. That is,

$$Q = \frac{F_T}{kY} = r^2 \left[\frac{1 + (2\zeta r)^2}{(1 - r^2)^2 + (2\zeta r)^2} \right] \tag{8.5}$$

The acceleration factor (AF) can be expressed as the product of the amplitude ratio of acceleration R and force transmissibility Q. That is,

$$\mathrm{AF} = \left(\frac{S_1}{S_0}\right)^n = \left(\frac{F_1}{F_0}\right)^\lambda = \left(\frac{a_1 F_T}{a_0 kY}\right)^\lambda = (R \times Q)^\lambda \tag{8.6}$$

For natural frequency ($r = 1.0$) and small damping ratio ($\zeta = 0.1$), the force transmissibility Q had a value of approximately 5.1 and the amplitude ratio of acceleration R was 4.17. Using a stress dependence of 2.0, the acceleration factor in Eq. (8.6) was found to be approximately 452.0 (Table 8.2).

8.2 Residential Sized Refrigerators During Transportation

(a) Failed locations in the field

(b) Failed connecting tubes in the mechanical compartment

Fig. 8.11 Fracture of the refrigerator connecting tubes in the field

Suppose that the shape parameter was 6.41 based on field data and the allowed failed numbers r was 0, the test time and the number of samples from Eq. (7.36) would be 40 min and 3 pieces for the first ALT. To meet the reliability target B1, there needs to be no fractured sample at the connecting tube of the refrigerator in 40 min that might be the Reliability quantitative (RQ) test specifications (Fig. 8.14).

For the first ALT, the connecting tubes in the mechanical compartments of three samples at 20 min were fracturing and the compressor rubber mounts were tearing during x-axis vibration tests. The estimated lifetime L_{B1} was approximately 3 days and estimated failure rate of the design samples λ was 2.9%/day. The shape and location of the failure in the ALT were similar to those seen in the field.

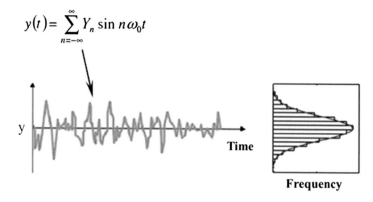

(a) A refrigerator subjected to base random vibrations

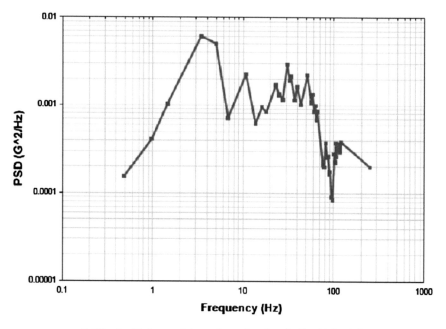

(b) Typical intermodal random vibration in the United States

Fig. 8.12 Refrigerators subjected to base random vibrations and their power spectral density

The reduction factor R also is 0.013 from the acceleration factor = 452 and shape parameter = 6.13. Consequently, we know that this parameter ALT is effective to save the testing time and sample size (Fig. 8.15).

The modified design parameters for the compressor compartment (or module) was modified as follows: (1) the shape of compressor rubber mount (C1: gap

8.2 Residential Sized Refrigerators During Transportation

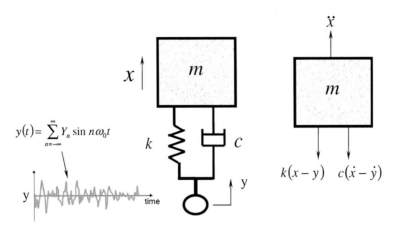

Fig. 8.13 A simplified model of the refrigerator subjected to repetitive random vibrations

Table 8.2 ALT conditions in refrigerator

System conditions	Worst case	ALT	AF
Transmissibility, Q ($r = 1.0$, $\zeta = 0.1$)	–	5.1 (from Eq. 8.5)	5.1①
Amplitude ratio of acceleration, R (a_1/a_0)	0.24 g	1 g	4.17②
Total AF (=(① × ②)^2)			452

reduction, 1.2 → 0.5 mm), (2) the shape of the connecting tube design (C2) (Fig. 8.16).

With these modified parameters, a second ALT was carried out and there were no problem at 40 and 60 min. The estimated lifetime L_{B1} was more than 7 days and the estimated failure rate of the design samples λ was less than 0.14%/day. Over the course of the two ALTs, refrigerators with the targeted B1 life were expected to survive without failure during cross country rail transport in the US. Table 8.3 shows a summary of the results of the ALTs, respectively.

8.3 Water Dispenser Lever in a Refrigerator

Figure 8.17 shows the Bottom Mounted Freezer (BMF) refrigerator with the newly designed water dispenser that consists of the dispenser cover (1), spring (2), and dispenser lever (3). As the consumer presses the lever, the dispenser system will supply water. To properly work this function, the dispenser system needs to be designed to handle the operating conditions subjected to it by the consumers who purchase and use the BMF refrigerator (Figs. 8.17 and 8.18).

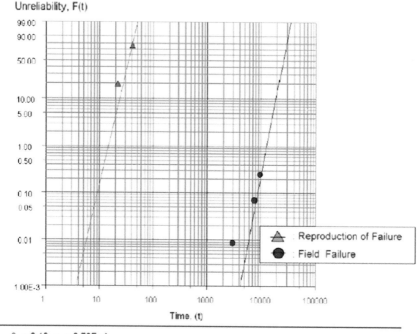

Fig. 8.14 Field data and results of accelerated life test on Weibull chart

In the field, the dispenser lever in the refrigerators had been fracturing, causing loss of the dispensing function. The field data on the failed products were important for understanding the use environment of consumers and helping to pinpoint design changes that needed to be made in the product. The dispenser system of a bottom-mounted refrigerator in field were cracking and fracturing under unknown consumer usage conditions. The damaged products might have had structural design flaws, including sharp corner angles resulting in stress risers in high stress areas. The design flaws combined with the repetitive loads on the dispenser lever could cause a crack to occur (Fig. 8.19).

The mechanical lever assembly of the water dispensing system consisted of many mechanical structural parts—the dispenser cover, spring, and dispenser lever. Depending on the consumer usage conditions, the lever assembly experienced repetitive mechanical loads in the water dispensing process. Figure 8.20 shows the functional design concept of the mechanical dispensing system. As a cup presses on the lever to dispense water, water will dispense. The number of water dispensing cycles will be influenced by consumer usage conditions. In the United States, the typical consumer requires a BMF refrigerator to dispense water from four up to 20 times a day.

8.3 Water Dispenser Lever in a Refrigerator

(a) Field   (b) 1st ALT Results

Fig. 8.15 Failure of refrigerator tubes in the field and 1st ALT result

Because the stress of the lever hinge depends on the applied force of the consumer, the life–stress model (LS model) can be modified as

$$\text{TF} = A(S)^{-n} = A(F)^{-\lambda} \qquad (8.7)$$

The acceleration factor (AF) can be derived as

$$\text{AF} = \left(\frac{S_1}{S_0}\right)^n = \left(\frac{F_1}{F_0}\right)^\lambda \qquad (8.8)$$

The dispenser is used on average 4–20 times per day. With a life cycle design point of 10 years, the dispenser incurs about 73,000 usage cycles. The applied force is 19.6 N which is the maximum force applied by the typical consumer.

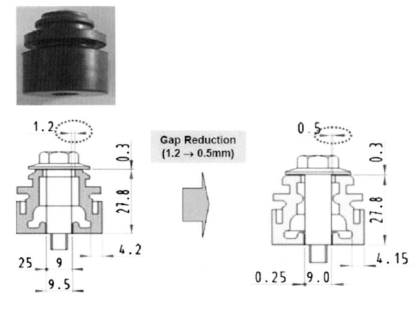

(a) Shape of compressor rubber (polymer) mount

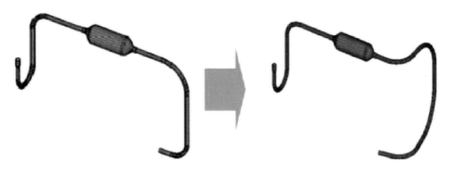

(b) Shape of the connecting tube design

Fig. 8.16 Modified design parameters of machine department (or module)

Doubling the applied force for the ALT to 39.2 N and using a stress dependence of 2.0, the acceleration factor is found to be approximately four in Eq. (8.8).

The test cycles and test sample calculated in Eq. (7.36) were 56,000 cycles and 8 pieces, respectively. The ALT was designed to ensure a B1 of 10 years life with about a 60% level of confidence that it would fail less than once during 56,000 cycles. Figure 8.21a shows the experimental setup of the ALT with labeled equipment for the robust design of the dispenser. Figure 8.21b shows the duty cycles for the pushing force F.

8.3 Water Dispenser Lever in a Refrigerator

Table 8.3 Results of ALT

	1st ALT	2nd ALT
	Initial design	Second design
In 45 min fracture of the connecting tube in refrigerator is less than 1	20 min: 1/3 fracture 40 min: 2/3 fracture	45 min: 3/3 OK 60 min: 3/3 OK
Machine room in refrigerator		
Material and spec	C1 shape of the compressor rubber C2 connecting tube design	

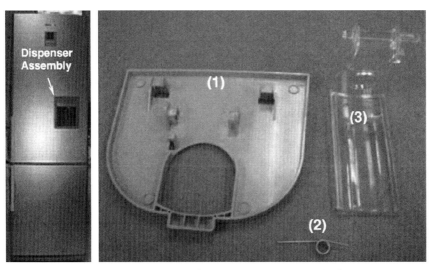

(a) BMF refrigerator (b) Mechanical parts of the dispenser lever assembly: Dispenser cover (1), Spring (2) and Dispenser lever (3)

Fig. 8.17 BMF refrigerator and dispenser assembly

Fig. 8.18 Robust design schematic of water dispensing

Key Noise Parameters
N1: Customer usage & load conditions
N2: Environmental conditions

Key Control Parameters
C1: Lever material & size

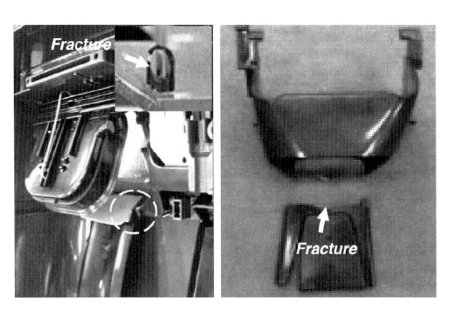

Fig. 8.19 A damaged product after use

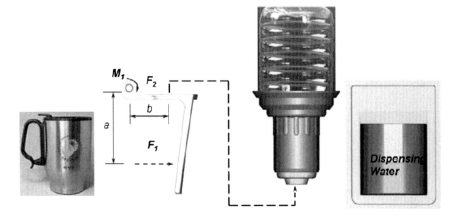

Fig. 8.20 Design concept of mechanical dispensing system

8.3 Water Dispenser Lever in a Refrigerator

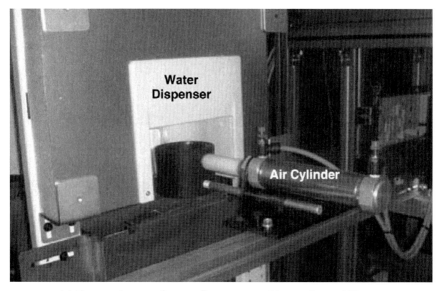

(a) Test equipment of water dispenser used in accelerated life testing

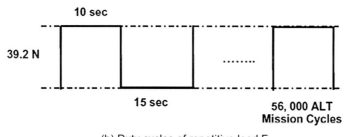

(b) Duty cycles of repetitive load F

Fig. 8.21 Equipment used in accelerated life testing and duty cycles of repetitive load F

An air cylinder controlled the pushing force, F of the cup. When the start button in the controller panel gave the start signal, the air cylinder with the mug-shape cup pressed the dispenser lever. At this point, the cup impacted the dispenser lever at the maximum mechanical force of 39.2 N.

Figure 8.22 shows the failed product from the field and from the accelerated life testing. In the photos, the shape and location of the failure in the ALT were similar to those seen in the field because of the stress raiser such as lever corner with no rounding. The reduction factor R also is 0.009 from the acceleration factor = 5 and shape parameter = 3.5. Consequently, we know that this parameter ALT is effective to save the testing time and sample size. These stress raisers in lever like no rounding should be improved to meet the reliability target of lever in the design phase.

Figure 8.23 also shows the photograph of the ALT results and field data and Weibull plot. The shape parameter in the first ALT was estimated at 2.0. For the

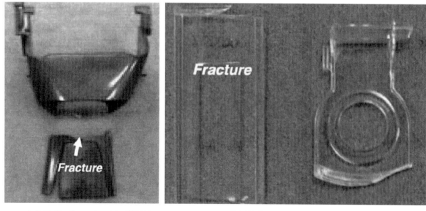

(a) Failed product in Field (b) Failed sample in ALT.

Fig. 8.22 Failed products in field and ALT

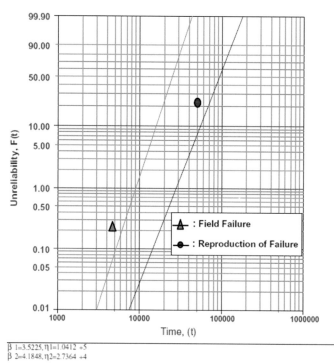

β 1=3.5225, η1=1.0412 +5
β 2=4.1848, η2=2.7364 +4

Fig. 8.23 Photograph of the ALT results and field data and Weibull plot

8.3 Water Dispenser Lever in a Refrigerator

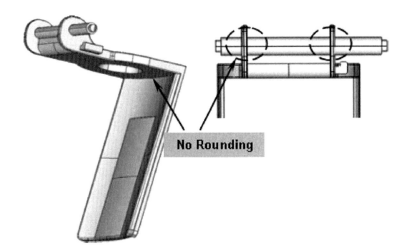

Fig. 8.24 Structure of failing dispenser lever in field

final design, the shape parameter was obtained from the Weibull plot and was determined to be 3.5. These methodologies were valid in pinpointing the weak designs responsible for failures in the field and supported by two findings in the data. The location and shape also, from the Weibull plot, the shape parameters of the ALT, $\beta1$, and market data, $\beta2$, were found to be similar.

The fracture of the dispenser lever in both the field products and the ALT test specimens occurred in both the front corner of the lever and the hinge (Fig. 8.24). The repetitive applied force in combination with the structural flaws may have caused cracking and fracture of the dispenser lever. The design flaw of sharp corners/angles resulting in stress risers in high stress areas can be corrected by implementing fillets on the hinge rib and front corner as well as increasing the hinge rib thickness. Through a finite element analysis, it was determined that the concentrated stresses resulting in fracture at the shaft hinge and the front corner were 8.37 and 5.66 MPa, respectively.

The confirmed values of AF and β in Fig. 8.23 were 4.0 and 3.5, respectively. The recalculated test cycles and sample size calculated in Eq. (7.36) were 56,000 and 8 EA, respectively. To meet the reliability target, three ALTs were performed to obtain the design parameters and their proper levels. In the three ALTs the dispenser lever cracked and/or fractured at the front corner of the lever and at the hinge in the first test, at the front corner of the lever in the second test, and at the front of the lever in the third test.

Table 8.4 shows the results of the design parameters confirmed from a tailored set of ALTs and a dispenser lever with high fatigue strength was redesigned by parametric ALTs (see Table 8.5). With these modified parameters, the BMF refrigerator can repetitively dispense water for a longer period without failure. Based on the modified design parameters, corrective measures taken to increase the life cycle of the dispenser system included: (1) increase the hinge rib rounding, C1,

Table 8.4 Redesigned dispenser lever

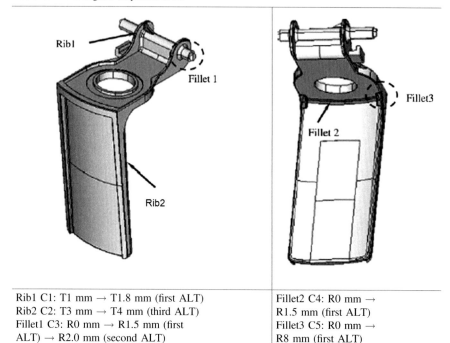

Rib1 C1: T1 mm → T1.8 mm (first ALT) Rib2 C2: T3 mm → T4 mm (third ALT) Fillet1 C3: R0 mm → R1.5 mm (first ALT) → R2.0 mm (second ALT)	Fillet2 C4: R0 mm → R1.5 mm (first ALT) Fillet3 C5: R0 mm → R8 mm (first ALT)

from 0.0 to 2.0 mm; (2) increase the front corner rounding, C2, from 0.0 to 1.5 mm; (3) increase the front side rounding, C3, from 0.0 to 11.0 mm; (4) increase the hinge rib thickness, C4, from 1.0 to 1.8 mm; and (5) increase the front lever thickness, C5, from 3.0 to 4.0 mm.

Figure 8.25 shows the graphical results of ALT plotted in a Weibull chart. Applying the new design parameters to the finite element analysis the stress concentrations in the shaft hinge decreased from 8.37 to 6.82 MPa and decreased in the front corner from 5.66 to 3.31 MPa. Over the course of the three ALTs the B1 life of the samples increased from 8.3 years to over 10.0 years.

8.4 Refrigerator Compressor Subjected to Repetitive Loads

A refrigerator consists of a compressor, a condenser, a capillary tube, and an evaporator. The refrigerant enters the compressor at a low pressure. It then leaves the compressor and enters the condenser at some elevated pressure; the refrigerant is condensed as heat is transferred to the surroundings. The refrigerant then leaves the condenser as a high-pressure liquid. The pressure of the liquid is decreased as it

8.4 Refrigerator Compressor Subjected to Repetitive Loads 243

Table 8.5 Results of ALT

	1st ALT	2nd ALT	3rd ALT
	Initial design	Second design	Final design
In 56,000 cycles, fracture of dispenser is less than 1	52,000 cycles: 2/8 fracture 74,000 cycles: 6/8 OK	56,000 cycles: 8/8 OK 67,500 cycles: 1/8 fracture 92,000 cycles: 7/8 OK	56,000 cycles: 8/8 OK 68,000 cycles: 1/8 fracture 92,000 cycles: 7/8 OK
Dispenser lever structure			
Material and specification	Rib1 T1.0 mm → T1.8 mm Fillet 1 R0.0 mm → R1.5 mm Fillet 2 R0.0 mm → R1.5 mm	Fillet 1 R1.5 mm → R2.0 mm Fillet 2 R8.0 mm → R11.0 mm	Rib2 T3.0 mm → T4.0 mm

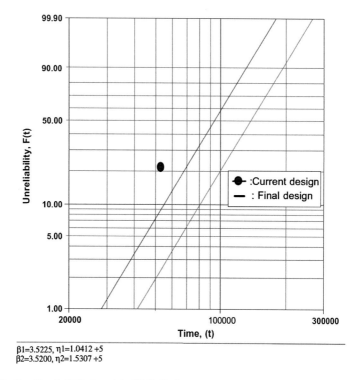

Fig. 8.25 Results of ALT plotted in Weibull chart

β1=3.5225, η1=1.0412 +5
β2=3.5200, η2=1.5307 +5

flows through the expansion valves, and as a result, some of the liquid flashes into cold vapor. The remaining liquid at a low pressure and temperature is vaporized in the evaporator as heat is transferred from the fresh/freezer compartment. This vapor then reenters the compressor. The main function of the refrigerator is to provide cold air from the evaporator to the freezer and refrigerator compartments (Fig. 8.26).

A capillary tube controls the flow in the refrigeration system and drops the high pressure of the refrigerant in the condenser to the low pressure in the evaporator. In a refrigeration cycle design, it is necessary to determine both the condensing pressure, P_c, and the evaporating pressure, P_e. These pressures depend on ambient conditions, customer usage conditions, and heat exchanger capacity in the initial design stage.

To derive the life–stress model and acceleration factor, the time-to-failure (TF) can be estimated from the McPherson's derivation:

$$\text{TF} = A(S)^{-n} \exp\left(\frac{E_a}{kT}\right) \qquad (8.9)$$

8.4 Refrigerator Compressor Subjected to Repetitive Loads

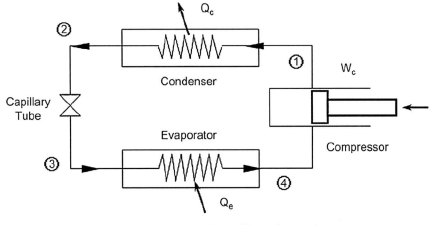

(a) A vapor-compression refrigeration cycle

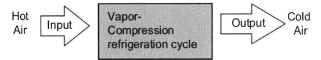

(b) Parameter diagram of refrigeration cycle

Fig. 8.26 A vapor-compression refrigeration cycle and its parameter diagram

To use Eq. (8.9) for accelerated testing, it needs to be modified and put into a more applicable form. A refrigeration system operates on the basic vapor-compression refrigeration cycle. The compressor receives refrigerant from the low-side (evaporator) and then compresses and transfers the refrigerant to the high-side (condenser) of the system. The capillary tube controls the flow in a refrigeration system and drops the high pressure of the refrigerant in the condenser to the low pressure in the evaporator. In a refrigeration cycle design, it is necessary to determine both the condensing pressure, P_c, and evaporating pressure, P_e, (see Fig. 8.26a).

The mass flow rate of refrigerant in a compressor can be modeled as

$$\dot{m} = \text{PD} \times \frac{\eta_v}{v_{\text{suc}}} \qquad (8.10)$$

The mass flow rate of refrigerant in a capillary tube can be modeled as

$$\dot{m}_{\text{cap}} = A \left[\frac{-\int_{P_2}^{P_3} \rho dP}{\frac{2}{D} f_m \Delta L + \ln\left(\frac{\rho_2}{\rho_3}\right)} \right]^{0.5} \qquad (8.11)$$

By conservation of mass, the mass flow rate can be determined as:

$$\dot{m} = \dot{m}_{\text{cap}} \qquad (8.12)$$

The energy balance in the condenser can be described as

$$Q_c = \dot{m}(h_1 - h_2) = (T_c - T_o)/R_c \qquad (8.13)$$

The energy balance in the evaporator can be described as

$$Q_c = \dot{m}(h_4 - h_3) = (T_i - T_e)/R_e \qquad (8.14)$$

When nonlinear Eq. (8.12) through (8.14) are solved, the mass flow rate, $\dot{m}$, evaporator temperature, T_e, and condenser temperature, T_c can be obtained. Because the saturation pressure, P_{sat}, is a function of temperature, the evaporator pressure, P_e (or condenser pressure P_c), can be obtained as:

$$P_e = f(T_e) \text{ or } P_c = f(T_c) \qquad (8.15)$$

One source of stress in a refrigeration system may come from the pressure difference between suction pressure, P_{suc}, and discharge pressure, P_{dis}.

For the theoretical single-stage cycle, the stress of the compressor depends on the pressure difference suction pressure, P_{suc}, and discharge pressure, P_{dis}. That is,

$$\Delta P = P_{\text{dis}} - P_{\text{suc}} \cong P_c - P_e \qquad (8.16)$$

By repeating the on and off cycles, the life of compressor shortens. The oil lubrication then relieves the stressful wear and extends the compressor life. Because the stress of the compressor depends on the pressure difference of the refrigerator cycle, the life–stress model can be modified as

$$\text{TF} = A(\Delta P)^{-n} \exp\left(\frac{E_a}{kT}\right) \qquad (8.17)$$

8.4 Refrigerator Compressor Subjected to Repetitive Loads

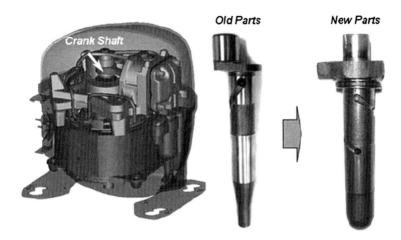

Fig. 8.27 Redesigned compressor and crankshaft

Figure 8.27 shows a redesigned crankshaft developed to reduce noise and improve energy efficiency of compressors in side-by-side (SBS) refrigerators. For these applications, the compressor needs to be designed robustly to operate under a wide range of customer usage conditions.

The acceleration factor (AF) from Eq. (8.17) can be derived as

$$\mathrm{AF} = \left(\frac{S_1}{S_0}\right)^n = \left(\frac{\Delta P_1}{\Delta P_0}\right)^\lambda \left[\frac{E_a}{k}\left(\frac{1}{T_0} - \frac{1}{T_1}\right)\right] \tag{8.18}$$

The normal number of operating cycles for one day was approximately ten; the worst case was twenty-four. Under the worst case, the objective compressor cycles for ten years would be 87,600 cycles.

The normal pressure was 1.07 MPa at 42 °C and the compressor dome temperature was 90 °C. It was measured after T type thermocouple pierced into the top compressor. For the accelerated testing, the acceleration factor (AF) for pressure at 1.96 MPa was 3.37 and for the compressor with a 120 °C dome temperature was 3.92 with a quotient, m, of 2. The total AF was approximately 13.2 (Table 8.6).

The parameter design criterion of the newly designed compressor can be more than the target life of B1 ten years. Assuming the shape parameter β was 1.9, the test cycles and test sample numbers calculated in Eq. (7.36) were 18,000 cycles and 30 pieces, respectively. The ALT was designed to ensure a B1 of ten years life with about a sixty-percent level of confidence that it would fail less than once during 18,000 cycles.

Figure 8.28 shows the ALT equipment used for the life testing in the laboratory. Figure 8.29 shows the duty cycles for the repetitive pressure difference, ΔP. For the ALT experiments, a simplified vapor-compression refrigeration system was fabricated. It consisted of an evaporator, compressor, condenser, and capillary tube.

Table 8.6 ALT conditions in a vapor-compression cycles

System conditions		Worst case	ALT	AF
Pressure (MPa)	High-side	1.07	1.96	3.36③
	Low-side	0.0	0.0	$(=(①/②)^2)$
	ΔP	1.07①	1.96②	
Temp. (°C)	Dome temp.	90	120	3.92④
Total AF (=③ × ④)		–		13.2

The inlet to the condenser section was at the top and the condenser outlet was at the bottom.

The condenser inlet was constructed with quick coupling and had a high-side pressure gage. A tengram refrigerator dryer was installed vertically at the condenser inlet. A thermal switch was attached to the condenser tubing at the top of the condenser coil to control the condenser fan. The evaporator inlet was at the bottom. At a location near the evaporator outlet, pressure gages were installed to enable access to the low-side for evacuation and refrigerant charging.

The condenser outlet was connected to the evaporator outlet with a capillary tube. The compressor was mounted on rubber pads and was connected to the condenser inlet and evaporator outlet. A fan and two 60 W lamps maintained the room temperature within an insulated (fiberglass) box. A thermal switch attached on the compressor top controlled a 51 m^3/h axial fan compressor, condenser, and capillary tube. The inlet to the condenser section was at the top and the condenser outlet.

In SBS units sold it was found that the crankshafts of some compressors were locking. Locking refers to the inability of the electric stator to rotate the crankshaft, due to a failure of one more component within the compressor under a range of unknown customer usage conditions. Field data indicated that the damaged products may have had a design flaw—oil lubrication problems. Due to this design flaw, the repetitive loads could create undue wear on the crankshaft and cause the compressor to lock.

Figure 8.30 shows the crankshaft of a locked-up compressor from the field and a sample from the accelerated life testing. In the photo, the shape and location of the parts in the failed product from the field were similar to those in the ALT results. Figure 8.31 represents the graphical analysis of the ALT results and field data on a Weibull plot. For the shape parameter, the estimated value in the previous ALT was 1.9. It was concluded that the methodologies used were valid in pinpointing the weaknesses in the original design of the units sold in the market because (1) the location and shape of the locking crankshaft from both the field and ALT were similar; and (2) on the Weilbull, the shape parameters of the ALT results, $\beta 1$ and market data, $\beta 2$, are very similar. The reduction factor R also is 0.15 from the acceleration factor = 13.2 and shape parameter = 1.9. Consequently, we know that this parameter ALT is effective to save the testing time and sample size.

When both the locked compressors from the field and the ALT compressor were cut apart, severe wear was found in regions of the crankshaft where there was no

8.4 Refrigerator Compressor Subjected to Repetitive Loads

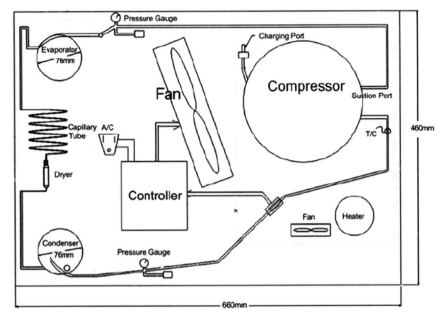

(a) A drawing of the test system

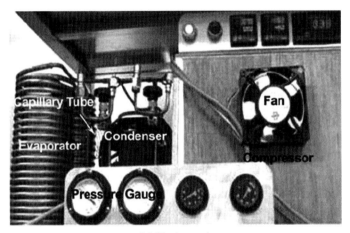

(b) Photograph

Fig. 8.28 Equipment used in accelerated life testing

Fig. 8.29 Duty cycles of repetitive pressure difference on the compressor

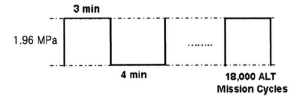

(a) Failed product in field   (b) Failed sample in 1st ALT

Fig. 8.30 Failed product in field and 1st ALT

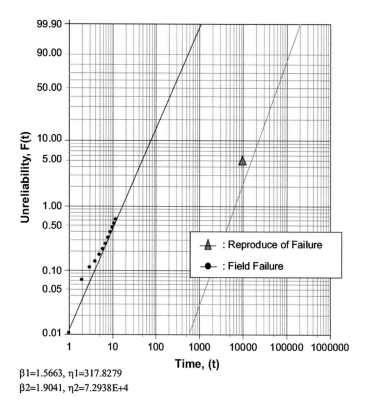

$\beta1=1.5663, \eta1=317.8279$
$\beta2=1.9041, \eta2=7.2938E+4$

Fig. 8.31 Field data and results of ALT on Weibull chart

8.4 Refrigerator Compressor Subjected to Repetitive Loads

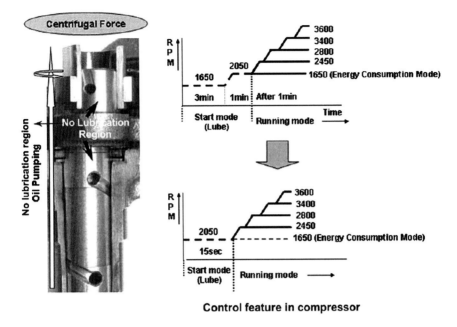

Fig. 8.32 No lubrication region in crankshaft and low starting RPM (1650 RPM)

lubrication—the friction area between shaft and connecting rod, and also the friction area between crankshaft and block. The locking of the compressor resulted from several design problems. There was (1) no oil lubrication in some regions of the crankshaft (Fig. 8.32); (2) a low starting RPM (1650 RPM) (Fig. 8.32); and (3) a crankshaft made from material with a wide range of hardness (FCD450) (Fig. 8.33).

The vital parameters in the design phase of the ALT were the lack of an oil lubrication region, low starting RPM, and weak crankshaft material. These compressor design flaws may cause the compressor to lock up suddenly when subjected to repetitive loads. The parameter design criterion of the newly designed samples was more than the target life, B1, of ten years. The confirmed values β on Weibull chart was 1.9. When the second ALT and third ALT proceeded, the recalculated test cycles and sample size calculated in Eq. (7.36) were 18,000 and 30 pieces, respectively. Based on the B1 life of ten years, the first, second, and third ALTs were performed to obtain the design parameters and proper levels. The compressor failure in the first ALT was due the compressor locking. In the second ALT, it was due to interference between the crankshaft and a thrust washer. During the third ALT, no problems were found with the compressor.

To improve the lubrication problems in the crankshaft, it was redesigned as the relocated lubrication holes, new groove, and new shaft material FCD500 (Fig. 8.34). To avoid the wear between crankshaft and washer, the minimum clearance was increased from 0.141 to 0.480 mm (Fig. 8.35). With these modified

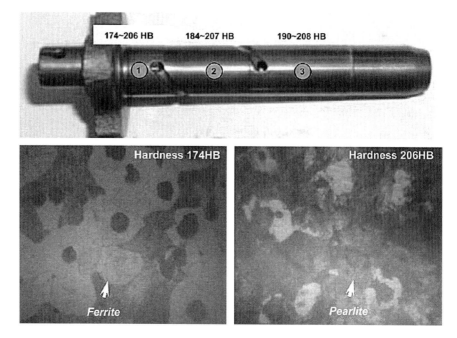

Fig. 8.33 A large variation of hardness (FCD450) in crankshaft

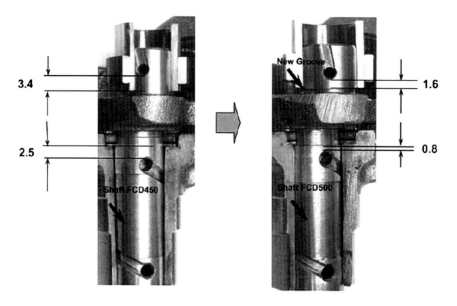

Fig. 8.34 Redesigned crankshaft in first ALT

8.4 Refrigerator Compressor Subjected to Repetitive Loads

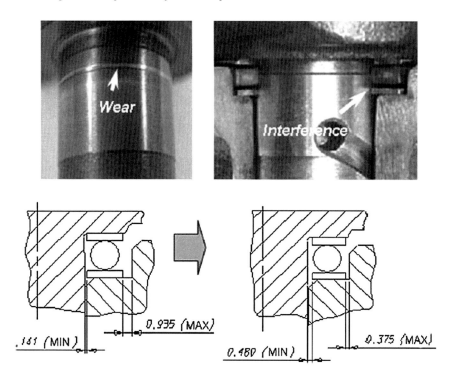

Fig. 8.35 Redesigned crankshaft in second ALT

design parameters, the SBS refrigerators can operate in the process of on and off repetitively with a B1 life of 10 years life.

The modified design parameters, with the corrective action plans, included (1) the modification of the oil lubrication region, C1; (2) increasing the starting RPM, C2, from 1650 to 2050; (3) changing the crankshaft material, C3, from FCD450 to FCD500; and (4) modifying the thrust washer dimension, C4, (see Table 8.7).

Table 8.7 provides a summary of the ALT results. Figure 8.36 show the results of ALT plotted in a Weibull chart. With the improved design parameters, the B1 life of the samples in the first, second, and third ALTs lengthen from 3.8 years to over 10.0 years.

8.5 Hinge Kit System (HKS) in a Kimchi Refrigerator

Figure 8.37 shows the Kimchi refrigerator with the newly designed hinge kit system. When a consumer closes the door, they want to close it conveniently and comfortable. For this function, the hinge kit system needs to be designed to handle the operating conditions subjected to it by the consumers who purchase and use the

Table 8.7 Results of ALT

	1st ALT	2nd ALT	3rd ALT
	Initial design	Second design	Final design
In 18,000 cycles, locking is less than 1	10,504 cycles: 2/30 locking 18,000 cycles: 28/30 OK	18,000 cycles: 2/30 wear 18,000 cycles: 28/30 OK	18,000 cycles: 30/30 OK 20,000 cycles: 30/30 OK
Crank shaft structure			
Material and specification	FCD450/FCD450 One new groove Location modification of oil supply holes	Modification of washer dimension	

8.5 Hinge Kit System (HKS) in a Kimchi Refrigerator

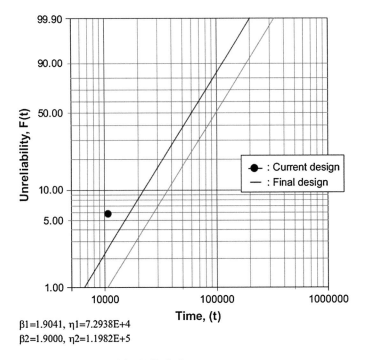

$\beta_1=1.9041$, $\eta_1=7.2938E+4$
$\beta_2=1.9000$, $\eta_2=1.1982E+5$

Fig. 8.36 Results of ALT plotted in Weibull chart

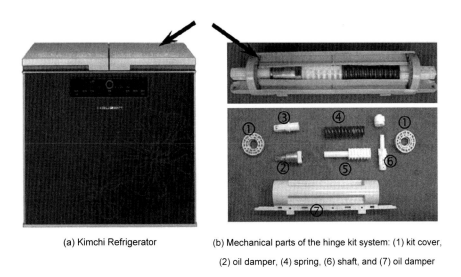

(a) Kimchi Refrigerator

(b) Mechanical parts of the hinge kit system: (1) kit cover, (2) oil damper, (4) spring, (6) shaft, and (7) oil damper

Fig. 8.37 Kimchi refrigerator and hinge kit assembly

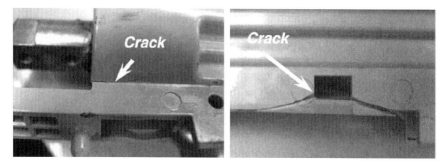

Fig. 8.38 Damaged products after use

Kimchi refrigerator. The hinge kit assembly consists of the kit cover, shaft, spring, oil damper, and kit housing, as shown in Fig. 8.37b. In the field, the hinge kit assembly in the refrigerators had been fracturing, causing the door not to close easily. Thus, the data on the failed products in the field were important for understanding the usage environment of consumers and helping to pinpoint design changes that needed to be made in the product.

In the field, parts of the hinge kit system of a Kimchi refrigerator were failing due to cracking and fracturing (Fig. 8.38) under unknown consumer usage conditions. Field data indicated that the damaged products might have had structural design flaws, including sharp corner angles and not enough enforced ribs resulting in stress risers in high stress areas. These design flaws combined with the repetitive loads on the hinge kit system could cause a crack to occur, and thus cause failure.

The mechanical hinge kit assembly of the door closing function consisted of many mechanical structural parts. Depending on the consumer usage conditions, the hinge kit assembly receives repetitive mechanical loads when the door is closed. Door closing involves two mechanical processes: (1) the consumer opens the door to take out the stored food, and (2) they then close the door by force.

Figure 8.39 shows the functional design concept of the mechanical hinge kit system in the accelerated testing. Figure 8.40 shows the robust design schematic

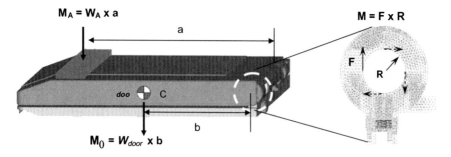

Fig. 8.39 Design concept of mechanical hinge kit system in the accelerated testing

8.5 Hinge Kit System (HKS) in a Kimchi Refrigerator

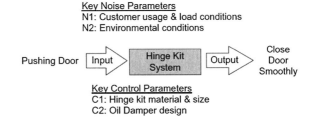

Fig. 8.40 Robust design schematic of hinge kit system

overview of the hinge kit system. As the consumer presses the refrigerator door, the hinge kit system helps to close the door smoothly. The stress due to the weight momentum of the door is concentrated on the hinge kit system. The number of door closing cycles will be influenced by consumer usage conditions. In the Korean domestic market, the typical consumer requires a Kimchi refrigerator the door system to open and close between three and ten times a day. The moment balance around the door system with an accelerated weight and the hinge kit system can be represented as

The moment balance around the HKS without accelerated weight in Fig. 8.39 can be represented as

$$M_0 = W_{\text{door}} \times b = T_0 = F_0 \times R \qquad (8.19)$$

The moment balance around the HKS with an accelerated weight can be represented as

$$M_1 = M_0 + M_A = W_{\text{door}} \times b + W_A \times a = T_1 = F_1 \times R \qquad (8.20)$$

Because F_0 is impact force in normal conditions and F_1 is impact force in accelerated weight, the stress on the HKS depends on the applied impact. Under the same temperature and efforts concept, the life–stress model (LS model) and can be modified as

$$\text{TF} = A(S)^{-n} = AT^{-n} = A(F \times R)^{-\lambda} \qquad (8.21)$$

The acceleration factor (AF) can be derived as

$$\text{AF} = \left(\frac{S_1}{S_0}\right)^n = \left(\frac{T_1}{T_0}\right)^\lambda = \left(\frac{F_1 \times R}{F_0 \times R}\right)^\lambda = \left(\frac{F_1}{F_0}\right)^\lambda \qquad (8.22)$$

The closing of the door occurs an estimated average 3–10 times per day. With a life cycle design point of 10 years, the hinge kit incurs about 36,500 usage cycles. For the worst case, the applied force around the hinge kit is 1.10 kN which is the maximum force applied by the typical consumer. The applied force for the ALT with accelerated weight is 2.76 kN. Using a stress dependence of 2.0, the acceleration factor is found to be approximately 6.3 in Eq. (8.22).

For the reliability target B1 of 10 years, the test cycles and test sample numbers calculated in Eq. (7.36) were 34,000 cycles and six pieces, respectively. The ALT was designed to ensure a B1 of ten years life with about a sixty-percent level of confidence that it would fail less than once during 34,000 cycles. Figure 8.41a shows the experimental setup of the ALT with labeled equipment for the robust design of the hinge kit system. Figure 8.41b shows the duty cycles for the impact force F.

The control panel on the top started or stopped the equipment, and indicated the completed test cycles and the test periods, such as sample on/off time. The door closing force F was controlled by the accelerated load applied to the door. When the start button in the controller panel gave the start signal, the simple hand-shaped arms held and lifted the Kimchi refrigerator door. At this point it impacted the hinge kit with the maximum mechanical impact force due to the accelerated weight (2.76 kN).

(a) Equipment used in accelerated life testing

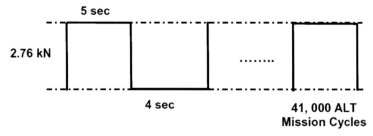

(b) Duty cycles of repetitive load F

Fig. 8.41 Equipment used in accelerated life testing and duty cycles of repetitive load F

8.5 Hinge Kit System (HKS) in a Kimchi Refrigerator

(a) Failed product from the field

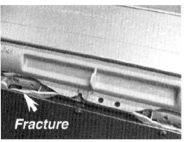
(b) Accelerated life testing

Fig. 8.42 Failed products in field and ALT

Figure 8.42 shows the failed product from the field and from the accelerated life testing, respectively. In the photos in Fig. 8.42, the shape and location of the failure in the ALT were similar to those seen in the field. Figure 8.43 represents the graphical analysis of the ALT results and field data on a Weibull plot. The shape parameter in the first ALT was estimated at 2.0. For the final design, the shape parameter was obtained from the Weibull plot and was determined to be 2.1.

These methodologies were valid in pinpointing the weak designs responsible for failures in the field and supported by two findings in the data. The location and shape from the Weibull plot, the shape parameters of the ALT ($\beta 1$) and market data ($\beta 2$) were found to be similar. The reduction factor R also is 0.016 from the experiment data—product lifetime, acceleration factor, actual mission cycles, and shape parameter. Consequently, we know that this parameter ALT is effective to decrease the testing time and sample size.

The fracture of the hinge kit in both the field products and the ALT test specimens occurred in the housing of the kit (Fig. 8.44a). The oil damper leaked oil in the hinge kit assembly (Fig. 8.44b). The repetitive applied force in combination with the structural flaws may have caused the fracturing of the hinge kit housing and the leak of the oil damper. The concentrated stresses of the housing hinge kit were approximately 21.2 MPa, based on finite element analysis. The stress risers in high stress areas resulted from the design flaws of sharp corners/angles, housing notches, and poorly enforced ribs.

The corrective action plan was to implement fillets, add the enforced ribs, and remove the notching on the housing of the hinge kit (Fig. 8.45). Applying the new design parameters to the finite element analysis, the stress concentrations in the housing of hinge kit decreased from 21.2 to 18.9 MPa.

The sealing structure of the oil damper had a 0.5 mm gap in the O-ring/Teflon/O-ring assembly. Due to the wear and impact, this sealing with the gap leaked easily. The sealing structure of the redesigned oil damper has no gap with Teflon/O-ring/Teflon (Fig. 8.46). The parameter design criterion of the newly designed samples was more than the target life of B1 of 10 years. The confirmed values of AF and β in Fig. 8.43 were 6.3 and 2.1, respectively. The test cycles and sample size recalculated in Eq. (7.36) were 41,000 and six pieces, respectively.

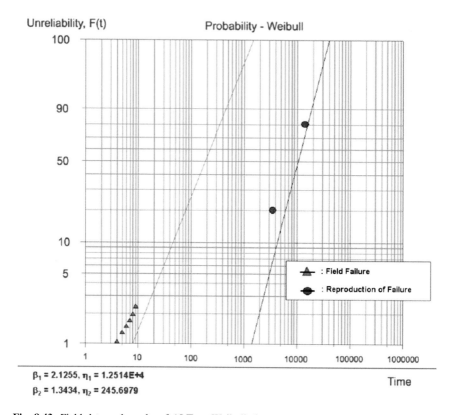

Fig. 8.43 Field data and results of ALT on Weibull chart

(a) The fracture of the hinge kit   (b) The leaked oil damper

Fig. 8.44 Structure of failing hinge kit system in accelerated testing

8.5 Hinge Kit System (HKS) in a Kimchi Refrigerator

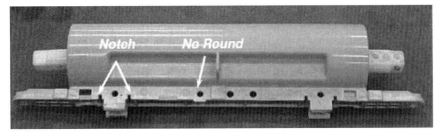

(a) Housing hinge kit structure

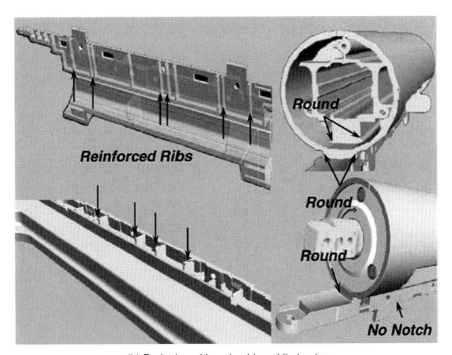

(b) Redesigned housing hinge kit structure

Fig. 8.45 Structure of newly designed hinge kit system

Based on the targeted BX and sample size, three ALTs were performed to obtain the design parameters and their proper levels. In the second ALTs the fracture of hinge kit cover occurs due to the repetitive impact stresses and its weak material. The cover housing of hinge kit assembly was modified by the material change from the plastics to the Al die-casting (Fig. 8.47).

The levels of the modified design parameters with corrective action plans were (1) the modification of the housing hinge kit (Fig. 8.45); (2) the modification of the oil sealing structure (see Fig. 8.46); (3) the material change of the cover housing (see Fig. 8.47).

262   8 Parametric ALT and Its Case Studies

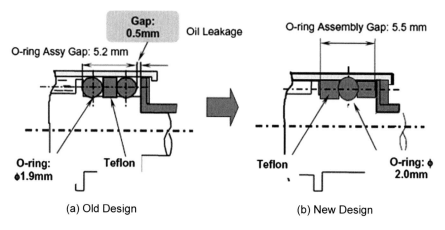

Fig. 8.46 Sealing structure of redesigned oil damper

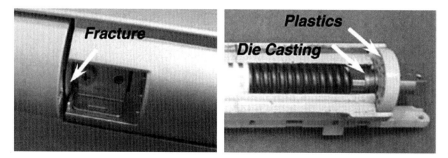

(a) Cover housing structure

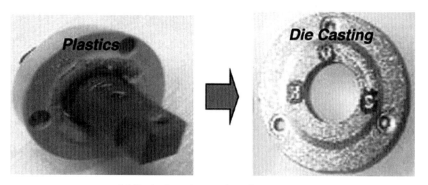

(b) Redesigned cover housing structure

Fig. 8.47 Redesigned cover housing structure

Table 8.8 shows the summary of the results of the ALTs, respectively. With these modified parameters, the Kimchi refrigerator can smoothly close the doors for a longer period without failure. Figure 8.48 shows the graphical results of the ALT plotted in a Weibull chart. Over the course of the three ALTs the B1 life of the samples increased from 8.3 years to over 10.0 years.

8.6 Refrigerator Drawer System

Figure 8.49 shows a refrigerator with the newly designed drawer and handle system and its parts. In the field, the refrigerator drawer and handle system had been failing, causing consumers to replace their refrigerators (Fig. 8.50). The specific causes of failures of the refrigerator drawers during operation were repetitive stress and/or the consumer improper usage. Field data indicated that the damaged products had structural design flaws, including sharp corner angles and weak ribs that resulted in stress risers in high stress areas.

A consumer stores food in a refrigerator to have convenient access to fresh food. Putting food in the refrigerator drawer involves opening the drawer to store or takeout food, closing the drawer by force. Depending on the consumer usage conditions, the drawer and handle parts receive repetitive mechanical loads when the consumer opens and closes the drawer.

Figure 8.51 shows the functional design concept of the drawer and handle system. The stress due to the weight load of the food is concentrated on the handle and support slide rail of the drawer. Thus, the drawer must be designed to endure these repetitive stresses. The force balance around the drawer and handle system can be expressed as:

Because the stress of the drawer and handle system depends on the food weight, the life–stress model (LS model) can be modified as follows:

$$\text{TF} = A(S)^{-n} = A(F_{\text{draw}})^{-\lambda} = A(\mu W_{\text{load}})^{-\lambda} \qquad (8.23)$$

The acceleration factor (AF) can be derived as

$$\text{AF} = \left(\frac{S_1}{S_0}\right)^n = \left(\frac{F_1}{F_0}\right)^{\lambda} = \left(\frac{\mu W_1}{\mu W_0}\right)^{\lambda} = \left(\frac{W_1}{W_0}\right)^{\lambda} \qquad (8.24)$$

The normal number of operating cycles for one day was approximately five; the worst case was nine. Under the worst case, the objective drawer open/close cycles for ten years would be 32,850 cycles. For the worst case, the food weight force on the handle of the drawer was 0.34 kN. The applied food weight force for the ALT was 0.68 kN. With a quotient, n, of 2, the total AF was approximately 4.0 using

Table 8.8 Results of ALT

	1st ALT	2nd ALT	3rd ALT
	Initial design	Second design	Final design
In 41,000 cycles, fracture is less than one	3340 cycles: 2/6 crack 15,000 cycles: 4/6 crack	7800 cycles: 1/6 crack 9200 cycles: 3/6 crack 14,000 cycles: 1/6 crack 26,200 cycles: 1/6 crack	41,000 cycles: 6/6 OK 74,000 cycles: 6/6 OK
Hinge kit structure	*Fracture*	*Fracture*	
Material and specification	C1: Redesigned housing hinge kit C2: Oil damper	C3: Plastic → Al die-casting	

8.6 Refrigerator Drawer System

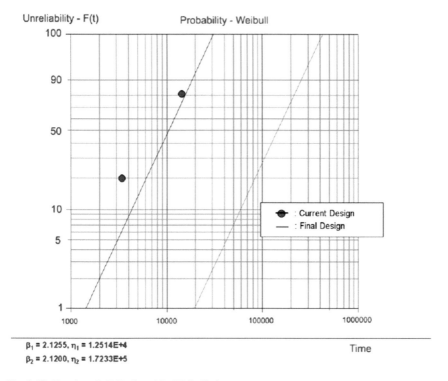

$\beta_1 = 2.1255, \eta_1 = 1.2514E+4$
$\beta_2 = 2.1200, \eta_2 = 1.7233E+5$

Fig. 8.48 Results of ALT plotted in Weibull chart

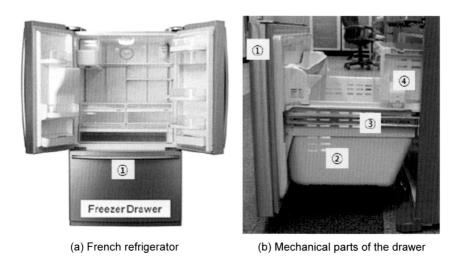

(a) French refrigerator　　　(b) Mechanical parts of the drawer

Fig. 8.49 Refrigerator and drawer assembly: handle ①, drawer ②, slide rail ③, and pocket box ④

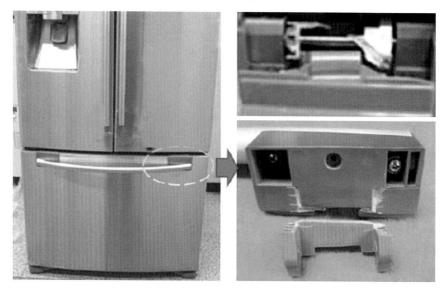

Fig. 8.50 A damaged product after use

Eq. (8.24). The parameter design criterion of the newly designed drawer can be more than the target life of B1 of 10 years. Assuming the shape parameter β was 2.0, the test cycles and test sample numbers calculated in Eq. (7.36) were 67,000 cycles and 3 pieces, respectively. The ALT was designed to ensure a B1 life of 10 years with about a 60% level of confidence that it would fail less than once during 67,000 cycles.

Figure 8.52 shows ALT equipment and duty cycles for the repetitive food weight force, F_{draw}. For the ALT experiments, the control panel on top of the testing equipment started and stopped the drawer during the mission cycles. The food load, F, was controlled by the accelerated weight load in the drawer storage. When a button on the control panel was pushed, mechanical arms and hands pushed and pulled the drawer.

The fracture of the drawer in the first and second ALTs occurred in the handle and slide rails (Figs. 8.53b and 8.54). These design flaws in the handle and slide rails can result in a fracture when the repetitive food load is applied. To prevent the fracture problem and release the repetitive stresses, the handle and slide rails were redesigned. The corrective action plan for the design parameters included: (1) increasing the width of the reinforced handle, C1, from 90 to 122 mm; (2) increasing the handle hooker size, C2, from 8 to 19 mm; (3) increasing the rail fastening screw number, C3, from 1 to 2; (4) adding an inner chamber and plastic material, C4, from HIPS to ABS; (5) thickening the boss, C5, from 2.0 to 3.0 mm; (6) adding a new support rib, C6 (Table 8.9).

8.6 Refrigerator Drawer System

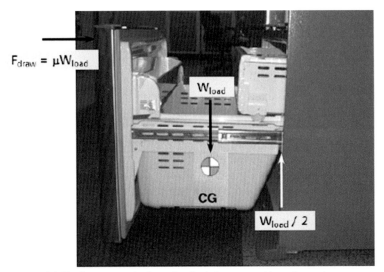

(a) Design concept of mechanical drawer and handle system

(b) Parameter diagram of drawer and handle system

Fig. 8.51 Functional design concept of the drawer and handle system

The parameter design criterion of the newly designed samples was more than the reliability target life, B1, of ten years. The confirmed value, β, on the Weibull chart in Fig. 8.55 was 3.1. The reduction factor R also is 0.0014 from the acceleration factor = 4 and shape parameter = 3.13. Consequently, we know that this parameter ALT is effective to save the testing time and sample size.

For the second ALT, the test cycles and sample size recalculated in Eq. (7.36) were 32,000 and 3 pieces, respectively. In the third ALT, no problems were found with the drawer after 32,000 cycles and 65,000 cycles. We therefore concluded that the modified design parameters were effective. Figure 8.56 shows the results of the 1st ALT and 3rd ALT plotted in a Weibull chart. Table 8.10 provides a summary of the ALT results. With the improved design parameters, B1 life of the samples in the third ALT was lengthened to more than 10.0 years.

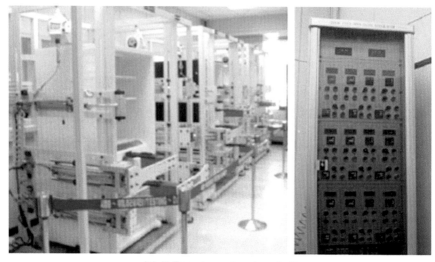

(a) ALT equipment and controller

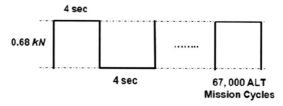

(b) Duty cycles of repetitive food weight force on the drawer

Fig. 8.52 ALT equipment and duty cycles

8.7 Compressor Suction Reed Valve

In the field, the suction reed valve in the compressor of the commercial refrigerator had been fracturing, causing loss of the cooling function (Fig. 8.57). The data on the failed products in the field were important for understanding how consumers used the refrigerators and pinpointing design changes that needed to be made to the product. The suction reed valves open and close to allow refrigerant to flow into the compressor during the intake cycle of the piston. Due to design flaws and repetitive stresses, the suction reed valves of domestic refrigerator compressors used in the field were cracking and fracturing, leading to failure of the compressor.

The fracture started in the void of the suction reed valve and propagated to the end (Fig. 8.58). Specific customer usage conditions and load patterns leading to the failures were unknown. Because the compressor locks up when the valve fails, the function of refrigerator is lost and customers would ask to have the refrigerator

8.7 Compressor Suction Reed Valve

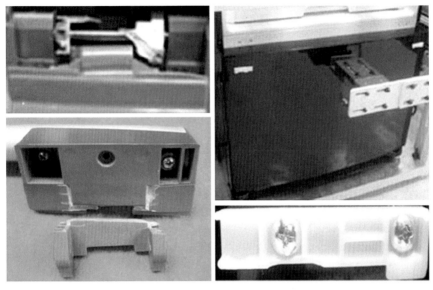

(a) Failed product in field (b) Failed sample in first accelerated life testing

Fig. 8.53 Failed products in field and first ALT

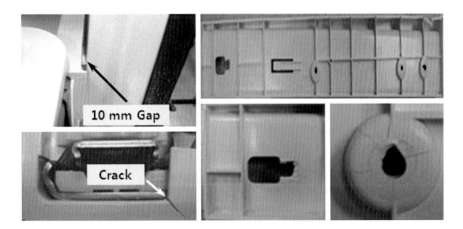

Fig. 8.54 Failed slide rails in second ALT

replaced. To solve this problem, it was very important to reproduce the field failure mode of the suction reed valve in the laboratory.

A refrigerator compressor assembly is a simple mechanical system that operates according to the basic principles of thermodynamics. The compressor receives refrigerant from the low-side (evaporator) and then compresses and transfers it to

Table 8.9 Redesigned handle and right/left slide rail

Handle	Right/left slide rail
C1: width L90 mm → L122 mm (1st ALT) C2: width L8 mm → L19 mm (1st ALT)	C3: Increase Screw 1 → 2 EA C4: Chamfer Rib Enforcement HIPS → ABS C5: Thickness Enforcement 2.0 → 3.0mm C6: Liner Support Enforcement Rib C3: Rail fastening screw number 1 → 2 (2nd ALT) C4: Chamfer: corner chamfer Plastic material HIPS → ABS (2nd ALT) C5: Boss thickness 2.0 → 3.0 mm (2nd ALT) C6: New support rib (2nd ALT)

8.7 Compressor Suction Reed Valve

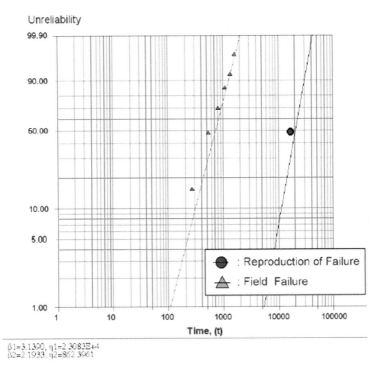

β1=3.1390, η1=2.3083E+4
β2=2.1933, η2=862.3961

Fig. 8.55 Field data and results of 1st ALT on Weibull chart

the high-side (condenser) of the system. Most compressor manufacturers are making every effort to develop more efficient, high-volume compressors. For these applications, the compressor needs to be designed robustly to operate under a wide range of customer usage conditions. The compressor assembly in the refrigerator in question consists of many mechanical parts, including the crankshaft, piston, valve plate (1), and suction reed valve (2) (see Figs. 8.59 and 8.60).

Analysis of the failed compressors from the field led to the postulate that there were two structural design flaws: (1) the suction reed valve had an overlap with the valve plate; and (2) the valve plate had a sharp edge (Fig. 8.61). When the suction reed valve impacted the valve plate over a long enough period of time, it would fracture.

The stress of the compressor depends on the pressure difference suction pressure, P_{suc}, and discharge pressure, P_{dis}. That is,

$$\Delta P = P_{\text{dis}} - P_{\text{suc}} \cong P_{\text{c}} - P_{\text{e}} \qquad (8.25)$$

For a refrigeration system, the time-to-failure, TF, can be modified as

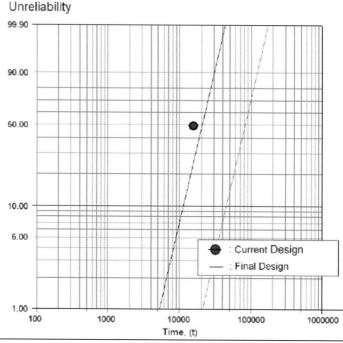

$\beta1=3.1390, \eta1=2.3083E+4$
$\beta2=3.1100, n2=9.2550E+4$

Fig. 8.56 Results of 1st ALT and 3rd ALT plotted in Weibull chart

$$\text{TF} = A(S)^{-n}\exp\left(\frac{E_a}{kT}\right) = A(\Delta P)^{-\lambda}\exp\left(\frac{E_a}{kT}\right) \quad (8.26)$$

The acceleration factor (AF) can be modified to include the load from Eq. (8.26):

$$\text{AF} = \left(\frac{S_1}{S_0}\right)^n \left[\frac{E_a}{k}\left(\frac{1}{T_0} - \frac{1}{T_1}\right)\right] = \left(\frac{\Delta P_1}{\Delta P_0}\right)^{\lambda} \left[\frac{E_a}{k}\left(\frac{1}{T_0} - \frac{1}{T_1}\right)\right] \quad (8.27)$$

The system was subjected to 22 on–off cycles per day under normal operating conditions. A worst case scenario was also simulated with 98 on–off cycles per day. Under the worst case conditions, the compressor operation for 10 years would be 357,700 cycles.

From the test data of the worst case, normal pressure was 1.27 MPa and the compressor dome temperature was 90 °C. For accelerated life testing, the acceleration factor (AF) for pressure was 2.94 MPa and the compressor dome temperature was 120 °C. With a quotient, n, of 2, the total AF was calculated using Eq. (8.26) to be 20.9 (see Table 8.11).

8.7 Compressor Suction Reed Valve

Table 8.10 Results of ALTs

	1st ALT	2nd ALT	3rd ALT
	Initial design	Second design	Final design
In 32,000 cycles, fracture is less than one	7500 cycles: 2/3 crack 12,000 cycles: 12,000 No problem	16,000 cycles: 2/3 crack	32,000 cycles: 3/3 OK 65,000 cycles: 3/3 OK
Hinge kit structure	![Fracture]	![10 mm Gap / Crack]	–
Material and specification	Width1: L90 → L122 Width2: L8 → L19.0	Rib1: new support rib Boss: 2.0 → 3.0 mm Chamfer1: corner material: HIPS → ABS	

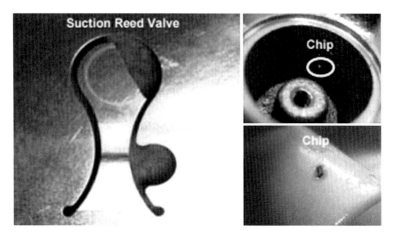

Fig. 8.57 Fracture of the compressor suction reed valve in the field

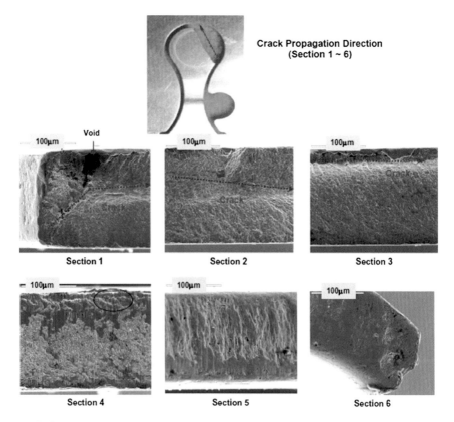

Fig. 8.58 Fractography of the compressor suction reed valve on SEM

8.7 Compressor Suction Reed Valve

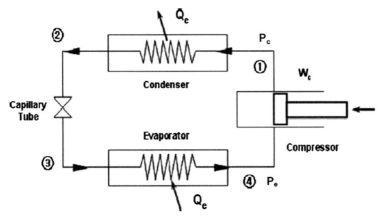

(a) Compressor system in a vapor-compression refrigeration cycle

(b) Parameter diagram of compressor system

Fig. 8.59 Schematic diagram for a compressor system

With a shape parameter, β, of 1.9, the test cycles and test sample numbers calculated in Eq. (7.36) were 40,000 cycles and 20 pieces, respectively. The ALT was designed to assure a B1 of 10 years with about a 60% level of confidence that no unit would fail during 40,000 cycles.

For the ALT experiments, a simplified vapor-compression refrigeration cycle was fabricated. It consisted of an evaporator, compressor, condenser, and capillary tube. A fan and two 60-W lamps maintained the temperature within the insulated (fiberglass) box. A thermal switch attached on the compressor top controlled a 51 m³/h axial fan. The test conditions and test limits were set up on the control board. As the test began, the high-side and low-side pressures could be observed on the pressure gage (see Fig. 8.62).

One sample in the first ALT ($n = 20$) failed after 8687 cycles. The confirmed value, β, based on field data was 1.9. The shapes and locations of the failures in

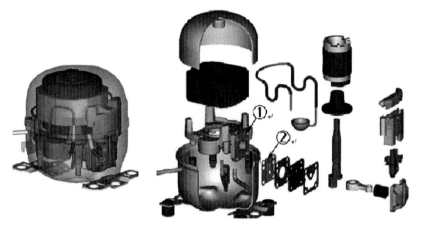

(a) Mechanical parts of the reciprocating compressor: crankshaft, piston, valve plate (1), suction reed valve (2)

(b) Valve plate (1) and suction reed valve (2)

Fig. 8.60 Reciprocating compressor and parts

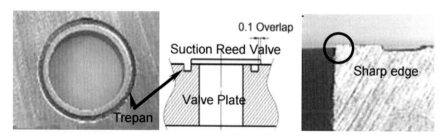

Fig. 8.61 Structure of suction reed and valve plate

8.7 Compressor Suction Reed Valve

Table 8.11 ALT conditions in a vapor-compression cycles

System conditions		Worst case, gage	ALT, gage	AF
Pressure (kg/cm^2)	High-side	13.0	30.0	5.3③ (=(①/②)^2)
	Low-side	0.0	0.0	
	ΔP	13①	30②	
Temp. (°C)	Dome temp.	90	120	3.92④
Total AF (=③ × ④)		–		20.9

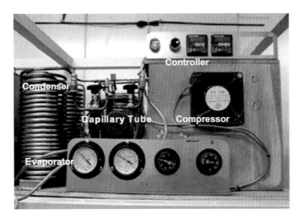

Fig. 8.62 Equipment for the compressor accelerated life tests

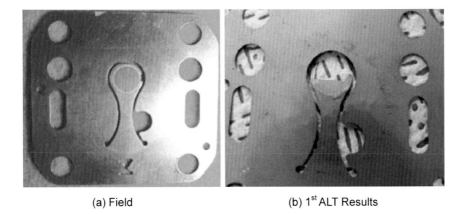

(a) Field   (b) 1st ALT Results

Fig. 8.63 Failure of suction reed valve in marketplace and first ALT result

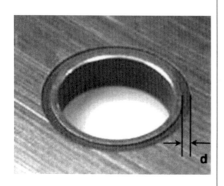

d = 0.73 mm → 1.25mm	SANDVIK 20C 0.178t (Carbon Steel)
Adding Ball Peening & & Brush Process	→SANDVIK 7C 0.178t (Stainless Steel)
	Adding tumbling process
(a) Valve plate	(b) Suction reed valve

Fig. 8.64 Redesigned suction reed and valve plate

samples from the first ALT and the field were similar (Fig. 8.63). The reduction factor R also is 0.2 from the acceleration factor = 20.9 and shape parameter = 1.89. Consequently, we know that this parameter ALT is effective to save the testing time and sample size.

The fracture of the suction reed valve came from its weak structure. It had the following characteristics: (1) an overlap with the valve plate; (2) weak material (0.178 t); and (3) a sharp edge on the valve plate, previously mentioned in Fig. 8.61.

When the suction reed valve impacted the valve plate continually, it suddenly fractures. The dominant failure mode of the compressor was leakage and locking due to the cracking and fracturing of the suction reed valve.

Figure 8.64 shows the redesigned suction reed valve and the valve plate. The valve controls the refrigerant gas during the process of suction and compression in the compressor. The suction reed valve required high bending/impact fatigue properties. The modified design parameters were: (1) increasing the trespan size of the valve plate from 0.73 to 1.25 mm, C1; (2) changing the material property from carbon steel (20C) to stainless steel (7C), C2; (3) adding a ball peening and tumbling process during the treating of suction reed valve, C3.

It would appear that the ALT methodology was valid for reproducing the failure found in the field. First, the location and shape of the fractured suction reed valves

8.7 Compressor Suction Reed Valve

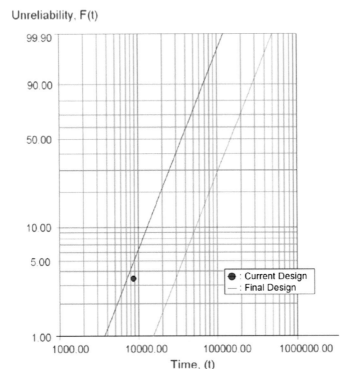

Fig. 8.65 Result of ALTs plotted in Weibull chart

from the field and those in the ALT results were similar. Figure 8.65 and Table 8.12 show the graphical results of an ALT plotted in a Weibull chart and the summary of the results of the ALTs, respectively.

8.8 Failure Analysis and Redesign of the Evaporator Tubing

Figure 8.66 shows the Kimchi refrigerator with the cooling aluminum evaporator tubing suggested for cost saving. When a consumer stores the food in the refrigerator, the refrigerant flows through the evaporator tubing in the cooling enclosure to maintain a constant temperature and preserves the freshness of the food. To

Table 8.12 Results of the ALTs

	1st ALT	2nd ALT
	Initial design	Second design
In 23,000 cycles, crack of suction reed valve is less than 1	8687 cycles: 1/20 8687 cycles: 19/20 stop	23,000 cycles: 60/60 OK 29,000 cycles: 60/60 OK
Suction reed valve and plate structure	<Suction reed valve>	<Plate plate>
Material and specification	C1: Trespan size d: 0.73 mm $\rightarrow$ 1.25 mm C2: Adding ball peening and rush process C3: SANDVIK 20C: 0.178t $\rightarrow$ 0.203t C4: Extending tumbling: 4 h $\rightarrow$ 14 h	

8.8 Failure Analysis and Redesign of the Evaporator Tubing

(a) Kimchi Refrigerator

(b) Mechanical parts of the hinge kit system: Inner Case (1), Evaporator tubing (2), Lokring (3), and Cotton adhesive tape (4)

Fig. 8.66 Kimchi refrigerator (**a**) and the cooling evaporator assembly (**b**)

perform this function, the tube in the evaporator need to be designed to reliably work under the operating conditions it is subjected to by the consumers who purchase and use the Kimchi refrigerator. The evaporator tube assembly in the cooling enclosure consists of an inner case (1), evaporator tubing (2), Lokring (3), and adhesive tape (4), as shown in Fig. 8.66b.

In the field, the evaporator tubing in the refrigerators had been pitting, causing loss of the refrigerant in the system, and resulting in the loss of cooling in the refrigerator. The data on the failed products in the field were important for understanding the usage environment of consumers and pinpointing design changes that needed to be made to the product (Fig. 8.67).

Field data indicated that the damaged products might have had design flaws. The design flaws combined with the repetitive loads could cause failure. The pitted surfaces of a failed specimen from the field were characterized by a scanning electron microscopy (SEM) and EDX spectrum (Fig. 8.68). We found a concentration of the chlorine in the pitted surface (Table 8.13). When Ion Liquid Chromatography (ILC) was used to measure the chlorine concentration, the result for the tubing having the cotton adhesive tape was 14 PPM. In contrast, the chlorine concentration for tubing having the generic transparent tape was 1.33 PPM. It was theorized that the high chlorine concentration found on the surface must have come from the cotton adhesive tape.

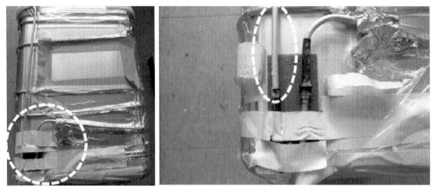

(a) Pitted evaporator tube in field

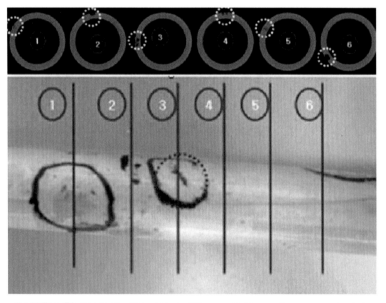

(b) X-Ray Photography showing a pitting corrosion on the evaporator tube

Fig. 8.67 A damaged product after use

As mentioned in Fig. 8.66, the evaporator tubing assembly in the cooling enclosure of the Kimchi refrigerator consists of many mechanical parts. Depending on the consumer usage conditions, the evaporator tubing experienced repetitive thermal duty loads due to the normal on/off cycling of the compressor to satisfy the thermal load in the refrigerator. Because the refrigerant temperatures are often below the dew point temperature of the air, condensation can form on the external surface of the tubing.

8.8 Failure Analysis and Redesign of the Evaporator Tubing

Fig. 8.68 SEM fractography showing a pitting corrosion on the evaporator tube

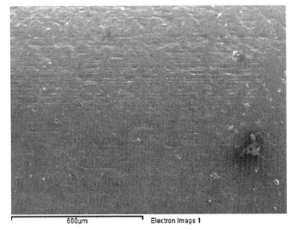

(a) No Pitting

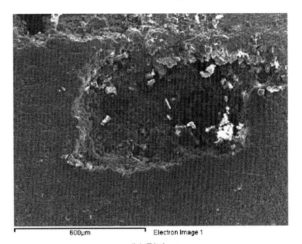

(b) Pitting

Table 8.13 Chemical composition of the no pitting and pitting surfaces

	No pitting		Pitting	
	Weight	Atomic	Weight	Atomic
O	11.95	18.65	25.82	37.38
Al	87.29	80.74	68.28	58.61
Cl	0.33	0.23	3.69	2.41
Si	0.42	0.38	0.66	0.55
Ca			0.70	0.40
K			0.50	0.30
Na			0.34	0.34
Totals	100.00		100.00	

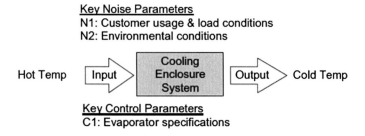

Fig. 8.69 Robust design schematic of a cooling enclosure system

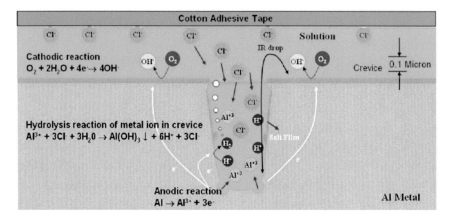

Fig. 8.70 An accelerating corrosion in the crevice due to low PH, high Cl⁻ concentration, depassivation and IR drop

Figure 8.69 shows a robust design schematic overview of the cooling evaporator system. Figure 8.70 shows the failure mechanism of the crevice (or pitting) corrosion that occurs because of the reaction between the cotton adhesive tape and the aluminum evaporator tubing. As a Kimchi refrigerator operates, water acts as an electrolyte and will condense between the cotton adhesive tape and the aluminum tubing. The crevice (or pitting) corrosion will begin.

The crevice (or pitting) corrosion mechanism on the aluminum evaporator tubing can be summarized as: (1) passive film breakdown by Cl⁻ attack; (2) rapid metal dissolution: $Al \rightarrow Al^{+3} + 3e^-$; (3) electro-migration of Cl into pit; (4) acidification by hydrolysis reaction: $Al^{+3} + 3H_2O \rightarrow Al(OH)_3 \downarrow + 3H^+$; (5) large cathode: external surface, small anode area: pit; and (6) the large voltage drop (i.e., "IR" drop, according to Ohm's Law $V = I \times R$, where R is the equivalent path resistance and I is the average current) between the pit and the external surface is the driving force for propagation of pitting.

The number of Kimchi refrigerator operation cycles is influenced by specific consumer usage conditions. In the Korean domestic market, the compressor can be

8.8 Failure Analysis and Redesign of the Evaporator Tubing

expected to cycle on and off 22–98 times a day to maintain the proper temperature inside the refrigerator.

Because the corrosion stress of the evaporator tubing depends on the corrosive load (F) that can be expressed as the concentration of the chlorine, the life–stress model (LS model) can be modified as

$$\text{TF} = A(S)^{-n} = A(F)^{-\lambda} = A(\text{Cl}\%)^{-\lambda} \tag{8.28}$$

The acceleration factor (AF) can be derived as

$$\text{AF} = \left(\frac{S_1}{S_0}\right)^n = \left(\frac{F_1}{F_0}\right)^\lambda = \left(\frac{\text{Cl}_1\%}{\text{Cl}_0\%}\right)^\lambda \tag{8.29}$$

The compressor in a Kimchi refrigerator is expected to cycle on average 22–98 times per day. With a life cycle design point of 10 years, the Kimchi refrigerator incurs 358,000 cycles. The chlorine concentration of the cotton adhesive tape was 14 PPM. To accelerate the pitting of the evaporator tubing, the chlorine concentration of the cotton tape was adjusted to approximately 140 PPM by adding some salt. Using a stress dependence of 2.0, the acceleration factor was found to be approximately 100 in Eq. (8.27).

For B1 life of 10 years, the test cycles and test sample numbers with the shape parameter $\beta = 6.41$ calculated in Eq. (7.36) were 4700 cycles and 18 pieces, respectively. The ALT was designed to ensure a B1 of 10 years life with about a sixty-percent level of confidence that it would fail less than once during 4700 cycles. Figure 8.71a shows the Kimchi refrigerators in accelerated life testing and an evaporator tubing in the enclosure contained a 0.2 M NaCl water solution. Figure 8.71b shows the duty cycles for the corrosive force (F) due to the chlorine concentration.

Figure 8.72 shows the failed product from the field and from the accelerated life testing, respectively. In the photos, the shape and location of the failure in the ALT were similar to those seen in the field. Figure 8.73 shows a graphical analysis of the ALT results and field data on a Weibull plot. These methodologies were valid in pinpointing the weak designs responsible for failures in the field and were supported by two findings in the data. The location and shape also, from the Weibull plot, the shape parameters of the ALT, ($\beta 1$), and market data, ($\beta 2$), were found to be similar.

The pitting of the evaporator tubing in both the field products and the ALT test specimens occurred in the inlet/outlet of the evaporator tubing (Fig. 8.74). Based on the modified design parameters, corrective measures taken to increase the life cycle of the evaporator tubing system included: (1) extending the length of the contraction tube (C1) from 50.0 to 200.0 mm; (2) replacing the cotton adhesive tape (C2) with the generic transparent tape.

Figure 8.75 shows a redesigned evaporator tubing with high corrosive fatigue strength. The confirmed values of AF and β in Fig. 8.73 were 100.0 and 6.41,

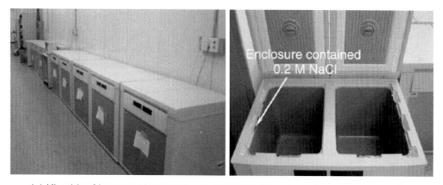

(a) Kimchi refrigerators in testing with 0.2 M NaCl water solution on evaporator

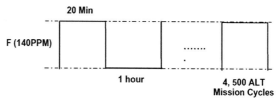

(b) Duty cycles of repetitive corrosive load F

Fig. 8.71 Kimchi refrigerators in accelerated life testing and duty cycles of repetitive corrosive load F

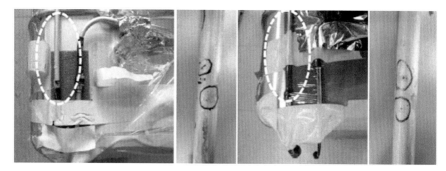

(a) Failed product from the field   (b) Accelerated life testing

Fig. 8.72 Failed products in field and ALT

respectively. The test cycles and sample size recalculated in Eq. (7.36) were 5300 and 8 EA, respectively. Based on the target BX life, two ALTs were performed to obtain the design parameters and their proper levels. In the two ALTs the outlet of the evaporator tubing was pitted in the first test and was not pitted in the second test.

8.8 Failure Analysis and Redesign of the Evaporator Tubing

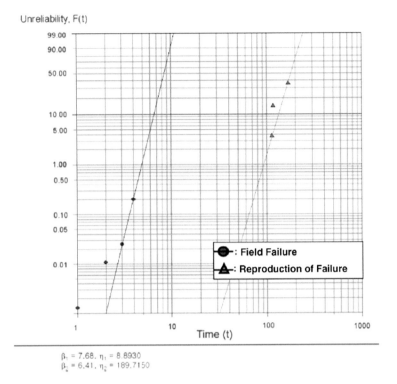

$\beta_1 = 7.68, \eta_1 = 8.8930$
$\beta_2 = 6.41, \eta_2 = 189.7150$

Fig. 8.73 Field data and results of ALT on Weibull chart

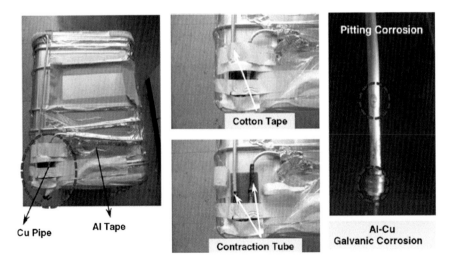

Fig. 8.74 Structure of pitting the corrosion tubing in field and the ALT test specimens

Fig. 8.75 A redesigned evaporator tubing

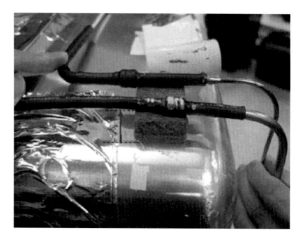

The repetitive corrosive force in combination with the high chlorine concentration of the cotton tape and the crevice between the cotton adhesive tape and the evaporator tubing contained the condensed water as an electrolyte may have been pitting.

With these modified parameters, the Kimchi refrigerator can reserve the food for a longer period without failure. Figure 8.76 and Table 8.14 show the graphical results of ALT plotted in a Weibull chart and the summary of the results of the ALTs, respectively. Over the course of the two ALTs the B1 life of the samples increased by over 10.0 years.

8.9 Compressor with Redesigned Rotor and Stator

A refrigerator system, which operates using the basic principles of thermodynamics, consists of a compressor, a condenser, a capillary tube, and an evaporator. The vapor-compression refrigeration cycle receives work from the compressor and transfers heat from the evaporator to the condenser. The main function of the refrigerator is to provide cold air from the evaporator to the freezer and refrigerator compartments. Consequently, it keeps the stored food fresh.

To improve its energy efficiency, designer would choose the good performance of compressor. Figure 8.77 shows a reciprocating compressor with redesigned rotor and stator. The redesign was developed to improve the energy efficiency and reduce the noise from the compressors in a side-by-side (SBS) refrigerator. For these applications, the compressor needed to be designed robustly to operate under a wide range of consumer usage conditions (Fig. 8.78).

As seen in Fig. 8.79, the reciprocating compressor in the refrigerators had been making noise in the field, causing the consumer to request replacement of their refrigerator. One of the specific causes of compressor failure during operation was

8.9 Compressor with Redesigned Rotor and Stator

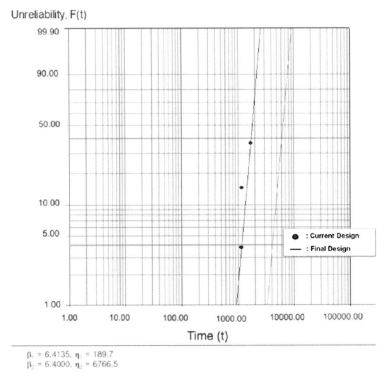

Fig. 8.76 Results of ALT plotted in Weibull chart

the compressor suspension spring. When the sound level during compressor shutdown of problematic refrigerators in the field was recorded, the result was approximately 46 dB (6.2 sones). The design flaws of the suspension spring in the problematic compressor were the number of turns and the mounting spring diameter. When the compressor would stop suddenly, the spring sometimes would not grab the stator frame tightly and would cause the noise.

After identifying the missing control parameters related to the newly designed compressor system, it was important to modify the defective compressor either through redesign of components or change the material used in the components. Failure analysis of marketplace data and accelerated life testing (ALT) can help to confirm the missing key control parameters and their levels in a newly designed compressor system.

In a refrigeration cycle design, it is necessary to determine both the condensing pressure, P_c, and the evaporating pressure, P_e. One indicator of the internal stresses on components in a compressor depends on the pressure difference between suction pressure, P_{suc}, and discharge pressure, P_{dis}, previously mentioned in Eq. (8.16). The acceleration factor (AF) can be derived as,

Table 8.14 Results of ALT

	1st ALT	2nd ALT
	Initial design	Second design
In 5300 cycles, corrosion of evaporator pipe is less than 1	1130 cycles: 1/18 pitting 1160 cycles: 2/18 pitting 1680 cycles: 4/18 pitting 1680 cycles: 11/18 OK	5300 cycles: 8/8 OK
Evaporator pipe structure		
Material and specification	Length of the contraction tube C1: 50.0 mm → 200.0 mm Adhesive tape type C2: Cotton type → generic transparent tape	

8.9 Compressor with Redesigned Rotor and Stator

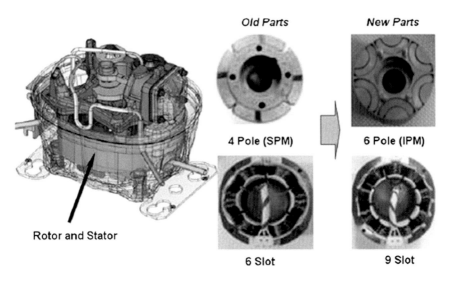

Fig. 8.77 Reciprocating compressors with redesigned rotor and stator

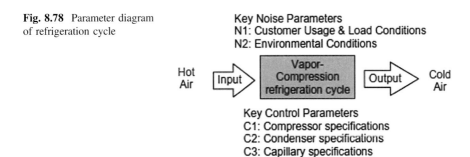

Fig. 8.78 Parameter diagram of refrigeration cycle

$$\text{AF} = \left(\frac{S_1}{S_0}\right)^n \left[\frac{E_a}{k}\left(\frac{1}{T_0} - \frac{1}{T_1}\right)\right] = \left(\frac{\Delta P_1}{\Delta P_0}\right)^\lambda \left[\frac{E_a}{k}\left(\frac{1}{T_0} - \frac{1}{T_1}\right)\right] \quad (8.30)$$

The normal number of operating cycles for 1 day was approximately 24; the worst case was 74. Under the worst case, the objective compressor cycles for 10 years would be 270,100 cycles. From the ASHRAE Handbook test data for R600a, the normal pressure was 0.40 MPa at 42 °C and the compressor dome temperature was 64 °C. For the accelerated testing, the acceleration factor (AF) for pressure at 1.96 MPa was 12.6 and for the compressor with a 110 °C dome temperature was 2.31 with a quotient, n, of 2. The total AF was approximately 29.2 (Table 8.15).

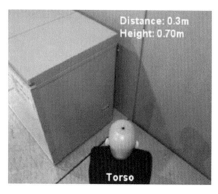
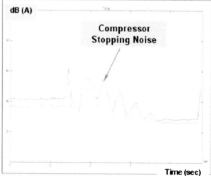

(a) Compressor stopping noise recorded with torso

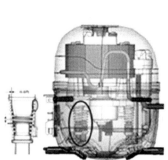

	Current	Modified
Suspension Spring		
Height, mm	30.2	31.2
Turn	7 / Upper 2 / Lower 2	7 / Upper 3 / Lower 2
Diameter, φ	15.55	15.35
Weight		6.9 kg

(b) Reciprocating compressor and the design flaws of suspension spring

Fig. 8.79 Stopping noise of the reciprocating compressor

The parameter design criterion of the newly designed compressor can be more than the target life of B1 10 years. Assuming the shape parameter β was 1.9, the test cycles and test sample numbers calculated in Eq. (7.36) were 9300 cycles and 100 pieces, respectively. The ALT was designed to ensure a B1 of 10 years life with about a 60% level of confidence that it would fail less than once during 9300 cycles. Figure 8.80a shows the duty cycles for the repetitive pressure difference ΔP.

Table 8.15 ALT conditions in a vapor-compression cycles for R600a

System conditions		Worst case	ALT	AF
Pressure (MPa)	High-side	0.40	1.39	12.6①
	Low-side	0.02	0.4	
	ΔP	0.38	1.35	
Temp. (°C)	Dome temp.	64	110	2.31②
Total AF (=① × ②)		–		29.3

8.9 Compressor with Redesigned Rotor and Stator

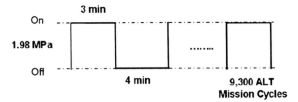

(a) Duty cycles of repetitive pressure difference on the compressor.

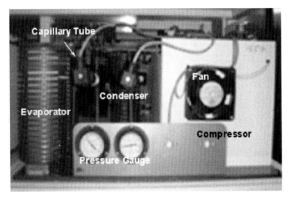

(b) Equipment used in Accelerated life testing.

Fig. 8.80 Duty cycles and equipment used in accelerated life testing

For the ALT experiments, a simplified vapor-compression refrigeration system was fabricated (see Fig. 8.80b).

Figure 8.81 shows the stopping noise and vibration of a compressor from the accelerated life testing. In the chart, the peak noise level and vibration of a normal sample in the compressor were 52 dB and 0.08 G when it stopped. On the other hand, for the failed sample #1, the peak noise levels and vibration were 65 dB and 0.52 G. For the failed sample #2, the peak noise levels and vibration were 70 dB and 0.60 G. Considering that the vibration specifications called for less than 0.2 G, the failed sample vibrations violated the specification. When the problematic samples in ALT equipment were mounted on the test refrigerator, the vibration was also reproduced with 0.25 G and violated the specification. In the field, consumer would request the failed samples to be replaced. Figure 8.82 represents the graphical analysis of the ALT results and field data on a Weibull plot. For the shape parameter, the estimated value on the chart was 1.9.

When the failed samples were cut apart, a scratch was found inside the upper shell of compressor where the stator frame had hit the shell. The gap between the frame and the shell was measured to be 2.9 mm. The design gap specification should have been more than 6 mm to avoid the compressor hitting the shell for the worst case. It was concluded that the stopping noise came from the hitting (or interference) between the stator frame and the upper shell. Thus, the tests pinpointed the design flaws in compressor (see Fig. 8.83a). For the shape parameter,

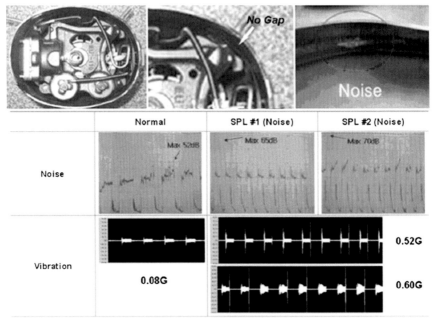

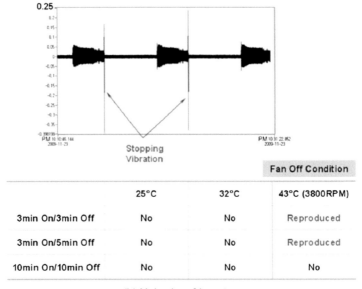

Fig. 8.81 Failed products in first ALT

8.9 Compressor with Redesigned Rotor and Stator

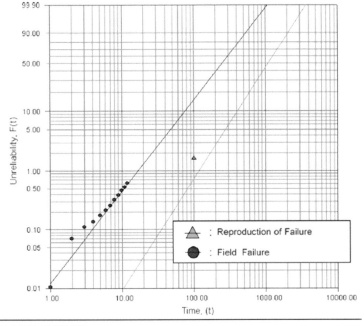

β1=1.5663, η1=317.8279
β2=1.8986, η2=1325.5404

Fig. 8.82 Field data and results of ALT on Weibull chart

the estimated value on the chart was 1.9 from the graphical analysis of the ALT results and field data on a Weibull plot. The vital missing parameter in the design phase of the ALT was a gap between the stator frame and the upper shell. These design flaws may make noise when the compressor stops suddenly. To reduce the noise problems in the frame, the shape of the stator frame were redesigned. As the test setup of the compressor assembly was modified to have more than a 6 mm gap, the gap size increased from 2.9 to 7.5 mm (Figs. 8.83b and 8.84).

The parameter design criterion of the newly designed samples was more than the target life, B1, of 10 years. The confirmed value, β, on the Weibull chart was 1.9. When the second ALT proceeded, the test cycles and sample size recalculated in Eq. (7.36) were 9300 and 100 pieces, respectively. In the second ALT, no problems were found with the compressor in 9300 cycles and 20,000 cycles. We expect that the modified design parameters are effective.

Table 8.16 provides a summary of the ALT results. Figure 8.85 shows the results of ALT plotted in a Weibull chart. With the improved design parameters, the B1 life of the samples in the second ALT lengthens more than 10.0 years.

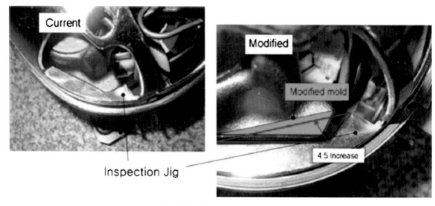

(a) Modified inspection jig

(Unit: mm)

	Frame & Shell GAP Spec. (6.0mm ↑)			
	A	B	C	D
As Is	8.5	9.1	3.2	2.9
To Be	8.3	8.8	7.7	7.5
Gap	0.2 ↓	0.3 ↓	4.5 ↑	4.6 ↑

(b) Gap between the stator frame and the upper shell

Fig. 8.83 Modified inspection jig and gaps

8.10 French Refrigerator Drawer System

Figure 8.86 shows the French refrigerator with the newly designed drawer system. When a consumer put food inside the refrigerator, they want to have convenient access to it and to have the food stay fresh. For this to occur, the draw system needs to be designed to withstand the operating conditions it is subjected to by the users. The drawer assembly consists of a box, left/right of the guide rail, and a support center, as shown in Fig. 8.86b.

In the field, parts of the drawer system of a French refrigerator were failing due to cracking and fracturing under unknown consumer usage conditions. Thus, the data on the failed products in the field were important for understanding the usage environment of consumers and helping to pinpoint design changes that needed to be made in the product (Fig. 8.87).

Field data indicated that the damaged products might have had structural design flaws, including sharp corner angles and weak ribs that resulted in stress risers in high stress areas. These design flaws that were combined with the repetitive loads on the drawer system could cause a crack to occur, and thus cause failure.

8.10 French Refrigerator Drawer System

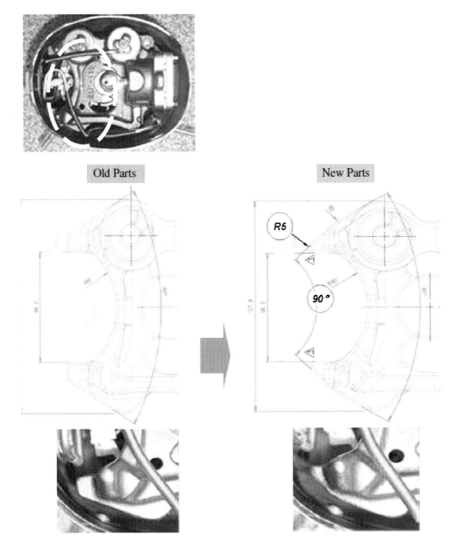

Fig. 8.84 Redesigned stator frame in second ALT

The drawer assembly consists of many mechanical structural parts. Depending on the consumer usage conditions, the drawer assembly receives repetitive mechanical loads when the drawer is opened and closed. Putting and storing food in the drawer involves two mechanical processes: (1) the consumer opens the drawer to store or take out the stored food, and (2) they then close the drawer by force.

Table 8.16 Results of ALT

	1st ALT	2nd ALT
	Initial design	Second design
In 9300 cycles, locking is less than 1	100 cycles: 2/100 noise 100 cycles: 98/100 OK	9300 cycles: 100/100 OK 20,000 cycles: 100/100 OK
Compressor structure	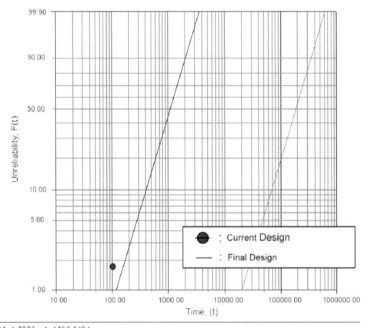	
Material and specification	C1: modification of the frame shape	

$\beta_1 = 1.8986, \eta_1 = 1325.5404$
$\beta_2 = 1.9000, \eta_2 = 2.2581E+5$

Fig. 8.85 Results of ALT plotted in Weibull chart

8.10 French Refrigerator Drawer System

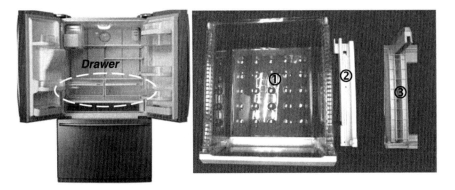

Fig. 8.86 French refrigerator and drawer assembly: vegetable box *1*, guide rail *2*, center support *3*

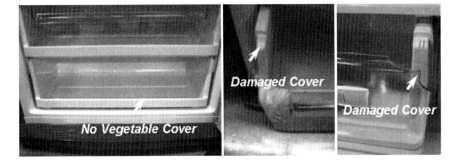

Fig. 8.87 A damaged products after use

Figure 8.88 shows the robust design schematic overview and the functional design concept of the drawer system. As the consumer stores the food, the drawer system helps to keep the food fresh. The stress due to the weight load of the food is concentrated on the drawer box and its support rails. And thus it is important to overcome these repetitive stresses when designing the drawer.

The number of drawer open and close cycles will be influenced by consumer usage conditions. In the United States, the typical consumer requires the drawer system of a French refrigerator to open and close between five and nine times a day.

The force balance around the drawer system can be represented as

$$F_{box} = \mu W_{load} \tag{8.31}$$

Because the stress of the drawer system depends on the applied force from the foods weight, the life–stress model (LS model) can be expressed as

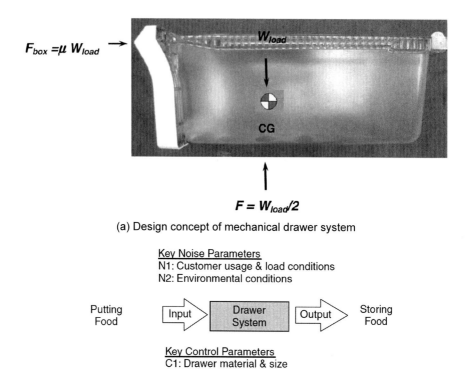

Fig. 8.88 Design concept and robust design schematic of mechanical drawer system

$$\text{TF} = A(S)^{-n} = A(F_{\text{box}})^{-\lambda} = A(\mu W_{\text{load}})^{-\lambda} \quad (8.32)$$

The acceleration factor (AF) can be derived as

$$\text{AF} = \left(\frac{S_1}{S_0}\right)^n = \left(\frac{F_1}{F_0}\right)^\lambda = \left(\frac{\mu W_1}{\mu W_0}\right)^\lambda = \left(\frac{W_1}{W_0}\right)^\lambda \quad (8.33)$$

The opening and closing of the drawer system occurs an estimated average five to nine times per day. With a life cycle design point of 10 years, the drawer would occur about 32,900 usage cycles. For the worst case, the weight force on the drawer is 0.59 kN which is the maximum force applied by the typical consumer. The applied weight force for the ALT was 1.17 kN. Using a stress dependence of 2.0, the acceleration factor was found to be approximately 4.0 using Eq. (8.33).

For B1 life, the test cycles and test sample numbers calculated in Eq. (7.36) were 22,000 cycles and six pieces, respectively. The ALT was designed to ensure a B1 life of 10 years life with about a 60% level of confidence that it would fail less than once during 22,000 cycles.

8.10 French Refrigerator Drawer System

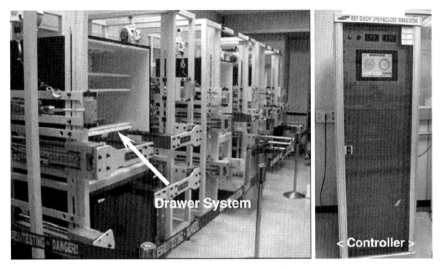

(a) Equipment used in accelerated life testing.

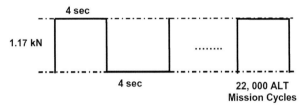

(b) Duty cycles of repetitive load F on the drawer system

Fig. 8.89 Equipment used in accelerated life testing and duty cycles

Figure 8.89 shows the experimental setup of the ALT with the test equipment and the duty cycles for the opening and closing force F. The control panel on the top of the testing equipment started and stopped the equipment, and indicated the completed test cycles and the test periods, such as sample on/off time. The drawer opening and closing force, F, was controlled by the accelerated weight load in the drawer system. When the start button in the controller panel gave the start signal, the simple hand-shaped arms held the drawer system. The arms then pushed and pulled the drawer with the accelerated weight force (1.17 kN).

Figure 8.90 shows the failed product from the field and from the accelerated life testing, respectively. In the photos, the shape and location of the failure in the ALT were similar to those seen in the field. Figure 8.91 represents the graphical analysis of the ALT results and field data on a Weibull plot.

The shape parameter in the first ALT was estimated at 2.0. For the final design, the shape parameter was obtained from the Weibull plot and was determined to be 3.6. These methodologies were valid in pinpointing the weak designs responsible

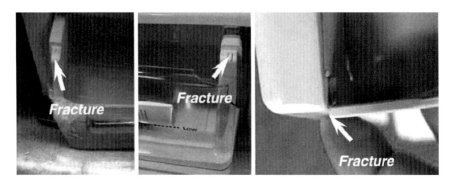

Fig. 8.90 Failed products in field (*left*) and 2nd ALT (*right*)

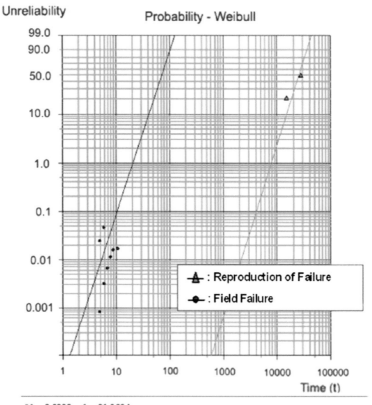

β1 = 3.3803, η1 = 81.0664
β2 = 3.5693, η2= 2.841E+4

Fig. 8.91 Field data and results of ALT on Weibull chart

8.10 French Refrigerator Drawer System

for failures in the field and were supported by two findings in the data. In the photo, the shape and location of the broken pieces in the failed market product are identical to those in the ALT results. And the shape parameters of the ALT ($\beta1$) and market data ($\beta2$) were found to be similar from the Weibull plot. The reduction factor R also is 0.034 from the acceleration factor = 4.0 and shape parameter = 1.9. Consequently, we know that this parameter ALT is effective to save the testing time and sample size.

Initially when the accelerated load of 12 kg was put into drawer, the center support rail was bent and the rollers on the left and right rail were broken away (Fig. 8.92). The design flaws of the bent center rail and the breakaway roller resulted in the drawers not sliding. The rail systems could be corrected by adding reinforced ribs on the center support rail as well as extruding the roller support to 7 mm (Table 8.17).

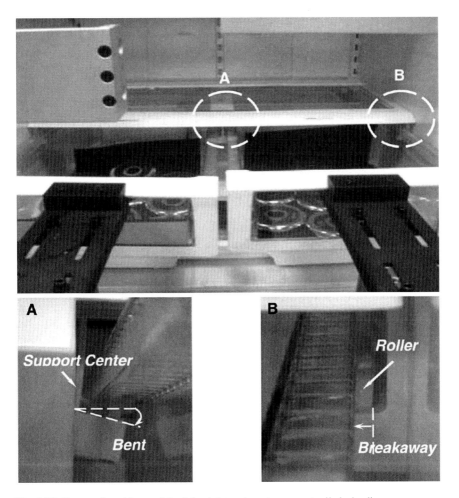

Fig. 8.92 Structural problems of the *left*, *right*, and *center* support rails in loading

Table 8.17 Redesigned box and center support rail

Box	Center support rail
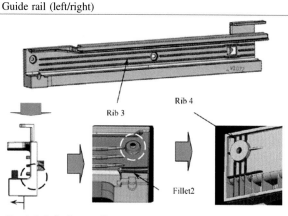	
C1: Rib1 T2.0 mm → T3.0 mm C2: Fillet R0.0 mm → R1.0 mm	C3: Rib2 new added rib C4: Extending rib1 L0.0 mm → L2.0 mm
Guide rail (left/right)	

Extrude 1: Roller 7mm outside

C5: Rib3 (new added rib, loading test)
C6: Extruder roller L0.0 mm → L7.0 mm
C7: Fillet R3 mm → R4 mm
C8: Rib4 New added back rib

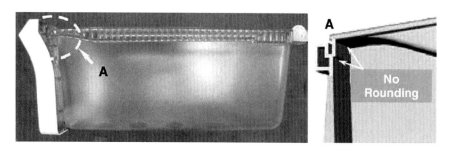

Fig. 8.93 Structure of failing drawer system in field

8.10 French Refrigerator Drawer System

Fig. 8.94 Structural problems of the left/right rail (*left*) and center support rails (*right*) in 1st ALT

Table 8.18 Results of ALT

	Initial design	Second design	Final design
In 22,000 cycles, fracturing is less than 1	3800 cycles: 3/6 fail 3800 cycles: 3/6 OK	15,000 cycles: 2/6 fail	28,000 cycles: 1/6 fail 28,000 cycles: 3/6 OK
Drawer structure			
Material and specification	Redesigned rail C1: Rib3 new added rib C2: Extrude1: L 0.0 mm → L 7.0 mm C3: Fillet2: R3 mm → R4 mm C4: Rib4: new added back	Redesigned box C5: Rib1 T2.0 mm → T3.0 mm C6: Fillet1 R0.0 mm → R1.0 mm	

The fracture of the drawer in both the field products and the ALT test specimens occurred in the intersection areas of the box and its cover (Fig. 8.93). The repetitive food loading forces in combination with the structural design flaws may have caused the fracturing of the drawer. The design flaws of no corner rounding and poorly enforced ribs resulted in the high stress areas. These flaws can be corrected by implementing the fillets and thickening the enforced ribs (Table 8.17).

The confirmed values of AF and β in Fig. 8.91 were 4.0 and 3.6, respectively. The test cycles and sample size recalculated in Eq. (7.36) were 22,000 and six pieces, respectively. Based on the targeted Bx, three ALTs were performed to obtain the design parameters and their proper levels. Due to repetitive i stresses, the

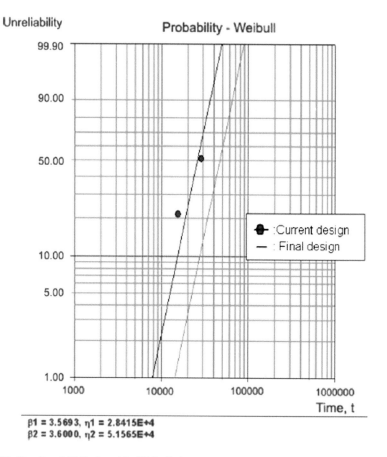

Fig. 8.95 Results of ALT plotted in Weibull chart

left and right rails of the drawer system cracked (Fig. 8.94a) and the roller of the support center sunk (Fig. 8.94b) in the first ALTs. Thus, a rib extruded 2 mm from the center support rail. And the left and light rail systems were corrected by design changes such as corner rounding and inserting ribs (Table 8.17).

Table 8.18 gives a summary of the results of the ALTs, respectively. Figure 8.95 shows the results of ALT plotted in a Weibull chart. With these modified parameters, the French refrigerator can smoothly open and close the drawers for a longer period without failure.

Chapter 9
Parametric ALT: A Powerful Tool for Future Engineering Development

Abstract This chapter discusses the concept of system engineering. Mechanical product is developing under the principle of system engineering. Product reliability becomes one of the product requirements. So when mechanical system with the sophisticated technology is put into plan, product reliability in the established design process should be implemented with reliability methodology-like parameter ALT. If not, new product will face quality problems. To settle down them, company will have to pay the quality costs.

Keywords System engineering (SE)

Today, new products such as automobiles, construction equipment, machine tools, airplane, domestic appliance, and bridges were designed under the principle of system engineering (SE). SE is an interdisciplinary field of engineering on how complex engineering projects should be designed and managed over product life cycles. Issues such as reliability, logistics, coordination of different teams (requirement management), evaluation measurements, and different disciplines become more difficult when dealing with large but complex projects. In systems engineering all aspects of a system are considered, and integrated into a whole (Fig. 9.1).

Company also would like to survive the limitless competition through the new technology development. Because there are a lot of things in the design phase for short developing duration, products often have inherent design problems. Due to their reliability disasters, engineers have become a critical factor to consider reliability in designing the product. The reliability should consider factor early in the design phase. The basic question is how to consider the reliability concept in the established design process. The company should have new quantitative developing process that considers the reliability factor in parallel with the established design process. If not, company will confront numerous recalls in the market.

Reliability disasters might come from the faulty components that have the missing design parameters not concerned in the process of R&D. When subjected to the wearout stress or overstress under the end user operating or environmental

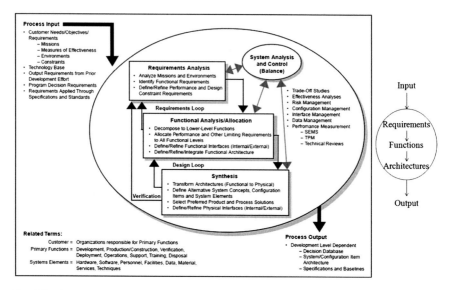

Fig. 9.1 The system engineering process [1]

conditions, the problematic components mounted in product cause failure. New product should be developed in the quantitative developing process that is included in (1) reliability target, (2) reliability testing and Weibull analysis, (3) design feedback, and (4) the analysis of the field failure data.

A new methodology for reliability design therefore is required to prevent the reliability disasters in the mechanical/civil system. The traditional qualitative methods—Capability maturity model integration (CMMI), FMEA, and FTA, are to look for the design problems on the documents. They only carry out to gather the design ideas or past experience by the representatives—planning, design, and production. Consequently, they often miss the chance to find a critical data in the design phase. The parametric ALT would be an alternative quantitative method to search out the missing design data because it uses the ALT plan, load analysis, and accelerated testing.

All mechanical products are fabricated from a multiple of structure to carry out the customer-required functions, which will tend to degrade or break down abruptly by random loads in the field. When mechanical/civil products are subjected to random loads, they would start the void in material (or design failures), propagate, and rupture it. If failure such as fatigue or fracture occurs, the product may no longer meet the required product functionality. To avoid failure, mechanical system should be designed to robustly withstand a variety of loads in a lifetime.

To accomplish the reliability design of modules in mechanical/civil product, the basic concepts of parametric ALT were discussed: (1) Setting overall parametric ALT plan of product; (2) Failure mechanics, design, and reliability testing; (3) Parametric accelerated life testing with an acceleration factor; and (4) Derivation of the sample size equation in Chap. 7. The failure modes and mechanisms of the

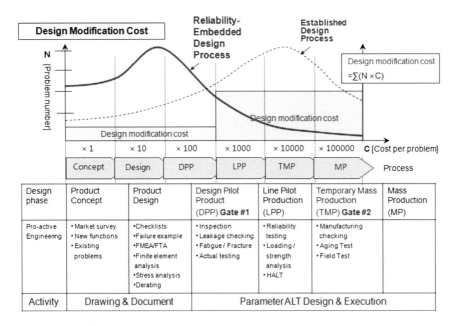

Fig. 9.2 Reliability-embedded design process

mechanical system in the field and parametric ALT may come from the missing design parameters or design flaws not considered in the design process. In the design phase the mechanical products should reveal the design flaws and establish action plans. To do it, the detail case studies on the design flaws were suggested in Chap. 8.

With the study of missing parameters in the design phase of the mechanical system, the parametric ALTs can be successful in proving a more reliable product or module with significantly longer life. The product or module with the modified design parameters will meet the reliability target. This reliability design methodologies will provide the reliability quantitative (RQ) test specifications of a mechanical structure that includes several assembly subjected to repetitive stresses under customer usage conditions. As a result, reliability-embedded design process will save the design modification cost because the problem number decreases (Fig. 9.2).

There are a variety of other structural systems—appliance, automobiles, airplane, machine tool, construction equipment, washing machines, and vacuum cleaners. For improving the reliability design of these systems, the missing controllable design parameters need to be identified to meet the targeted product (or module) reliability. And these principles of parametric ALT also are applicable to the area of civil engineering to design the construction structure. It is recommended that the missing controllable design parameters on these systems be further studied for reliability design of product in lifetime.

Reference

1. Defense of Department (2001) Systems engineering fundamentals. Defense Acquisition University Press, Virginia, p 31

Printed in the United States
By Bookmasters